T0104614

"I have become Space"

Thirteen Quantum and Twelve Cosmological Interpretations Awakened by a Novel Theory of Nature

Douglas W. Lipp

AuthorHouse™ LLC
1663 Liberty Drive
Bloomington, IN 47403
www.authorhouse.com
Phone: 1-800-839-8640

Published by AuthorHouse 04/22/2014

ISBN: 978-1-4969-0540-6 (sc)
ISBN: 978-1-4969-0541-3 (e)

Library of Congress Control Number: 2014907338

This book is printed on acid-free paper.

Table of Contents

Foreword

Douglas Lipp is and has been a good friend of mine for the past three decades. After a twenty year absence (mine) from Long Island I reunited with Doug on a South Shore beach by the name of Gilgo. We began our conversation without skipping a beat, catching up on physics and surfing, two mutual passions of ours. I had been in California (mostly San Diego) enjoying the surf and completing graduate school, eventually landing a job at the Salk Institute in La Jolla as a research technician, and an adjunct position in teaching basic Physics at the local community college. Doug had been living on Long Island in the bayside community of Babylon during my hiatus, and so it was natural for us to discuss Physics because simultaneously we had both been involved with it. Doug's genuine interest in and passion for matter, time, and space consumed his spare hours and I think was as much a hobby for him as was his surfing.

He might very well have spent a significant portion of his life doing theoretical physics and surfing exclusively given the opportunity and money to afford this, but fate intervened. He met his lovely wife and brought two beautiful boys into the world. Doug took a job at UL (Underwriters Laboratories), followed by another job at The Govmark Organization, the flammability laboratory where he currently works.

Doug has focused his attention on a theory that sees matter becoming space as that matter travels with our accelerating

Universe. More precisely, as matter travelling approaches "c", it will be transformed into space. I had the good fortune to be the recipient of one of Doug's generous gifts; a cap emblazoned with the letters M, T, and S, and that cap (insiders will know) really says it all regarding the "Coney Island Green Theory" . . . . Matter, over a period of Time spent travelling at relativistic speeds, will become Space.

Enjoy the book.

Edward Arikian

Preface

Welcome to, in the author's view, a paradigm shift.

The book you now hold serves up on the platter of science a new theory of nature. This new theory offers solutions and interpretations to many unanswered questions in the fields of Quantum Physics, the science of the very small, and Cosmology, the very large. The theory combines the two worlds. This theory I have named The Coney Island Green Theory, or as hereon in offered, CIG.

Some limited foundation in the key concepts within this book (i.e. Double Slit, Dark Matter) will be necessary to fully contemplate the books offered solutions to the noted conundrums facing science today. To make better sense of CIG's offerings, a read of Wikipedia on the noted topics may suffice in lieu of college coursework. Trained Physicists and Cosmologists should dig in and run deep with the ideas as your experience and depth of knowledge far exceeds my own, as far as applying CIG going forward. After all, you are the experts.

Please follow the thought process of these writings to discover CIG's interpretation of:

Red Shift
Dark Matter
Dark Energy

Quantum Gravity
Horizon Problem
Expanding Universe
Red Shift Anomalies
Accelerating Universe
The Double Slit Experiment (the particle becomes spatial and can go through both slits without the single point particle being in two places at once)

The physical reality of $E=mc^2$

The above, and more, in all their entirety are explained unambiguously via CIG's Synthesis of the Space-Time continuum (actually non-continuum) with the Mass-Energy equation ($E=mc^2$).

This is the MTS equation.

The theory offers the Quantification of Matter to Space, in terms of a new unit named the CUPI (pronounced QP, as in the doll, named in honor of my wife).

CIG theory is introduced as a relativistic theory offering new possibilities, and, where Albert Einstein stopped with matter warping the space-time continuum, CIG theory takes the next logical step and proves (actually nothing can ever be proven with pink elephants in the room) that it is the space-time continuum itself that actually turns into matter. All matter, inclusive of every particle in the Standard Model, is simply the degree of warp of space-time. Matter is made up of Space and Time. The theory will explain this.

Where there is a different time there must be a different place. Where there is a different place, there is a different space. Where there are different spaces, there are different volumes. CIG theory explains the creation of new volumes of space created as the result of different times imparted onto the world universe and as a direct result of the relativistic nature of nature arising from different rates of traveling massive particles. It is a new interpretation of time dilation and length contraction.

The book that firmly planted this author's theory was The Universe and Dr. Einstein by Lincoln Barnett, foreword by Albert Einstein—all this in the late nineteen seventies. The theoretical portion of MT=S started out from a simple but erred "rate multiplied by time is equal to distance" concept, from there, the speed of light multiplied by time is equal to distance; speed of light multiplied by time is equal to Space; speed of light (photon) multiplied by time is equal to Space; speed of light (photon/mass/infinite mass/matter) multiplied by Time is equal to Space; and finally Matter multiplied by Time is equal to Space, MT=S. The recognition of the correction to the erred beginning (so simply as namely for RT=D there needs to be a mass at travel . . . obviously) came a short time later. While the actual quantification of mass into a spatial volume came more recently.

The thinking parallels that of Kinetic Energy = 1/2 mv2, so all the Energy must equal all the mass (not just half of it) multiplied by the highest velocity possible, or that of light, for . . . . , $E=mc^2$! This would also mean that Potential Energy is also equal to 1/2mv2, which confirms that PE + KE = "E" = mc^2, thus maintaining appropriate symmetry. The above checks because any loss of KE to PE, can simultaneously be considered a loss of PE to KE, depending upon the view. But why digress this early. So let's not digress.

The CUPI represents that spatial volumetric quantity associated with one atomic mass unit and will be explained later.

As far as the Universal Expansion is concerned, matter is turning into space, and according to our mass to space conversion.

The Mathematical Equation: M x T = S, or MT = S, or Matter multiplied by Time is equal to Space explains the expanding universe. At times within the book the equivalent form of MTS is used.

The constructs of Matter are Space and Time.

The constructs of Space are Matter and Time.

The constructs of Time are Matter and Space:

Space is equal to Matter multiplied by Time

Matter is equal to Space divided by Time.

Time is equal to Space divided by Matter.

The Three Equations:

1. S=MT (Big Bang / New Bohr Orbitals) (forward "T" vector)
2. M=S/T (Big Crunch / Black Holes) (reverse "T" vector)
3. T=S/M (The Means, "T" as an arrow vector quantity)

Or, combined, simply: MTS

Where: M = Matter, T = Forward/Reverse Vector Time (% "c"), and S = Space

Note: "c" of course is the speed of light (not in a vacuum—it is the vacuum)

% "c" = the mass moving at its possible rates of travel (speeds): zero to "c" mph in the Reverse Vector Time direction and from "c" to zero mph in the Forward Vector Time direction. In this Biblical scheme of direction Time started with the void (S) and proceeded to M in the forward dirction of Time. This is the "c" to zero mph direction. In this manner, everything started with Light ["c"]. While we could then offer the STM equation, since most of us think in terms of matter before Space, we will keep MTS intact. Where MT = S appears in this book, recognize that it is in one direction and the other direction (S/T = M) can be combined as MTS.

At "c", mass manifests itself into a new spatial quantity, namely the pure vacuum. Mass follows Lorentz transformation percentages as it offers its equivalency of mass into space. From: Black Hole to Dark Matter to Dark Energy (zero % "c" to full "c")

The theory is not without a deep problematic philosophical consequence, namely that we can reduce matter to a simple permutation of space and time, such that when I walk down Main Street, I must avoid thinking of this nature, for to reduce matter to space and time, taking away all its particle-like material qualities, akin to removing fleshly matter itself, I literally find myself consisting of simply "Space". This is quite disturbing, perhaps much more so than the results of the Michelson/Morley experiment wherein the rate nature of light was found to be the same without regard to the aether flow, leading to the only explanation (at that time in science) of "contraction of length".

For it is this length contraction, heretofore not taken literally to mean that "Length unfolds into the dematerialization of Matter into Space", that the hat of the theory is hung upon, putting full value on its true meaning. The theory thus builds upon the scientific past. I should like to think that Einstein would have come to the same conclusion if at the time in 1905 he was aware of Hubble's red shift expansion. But, seeing a static universe was the only one known at the time, the Great Professor was not in a position to carry the events further.

So, for the purpose of being, I shall still consider myself as consisting of matter. That way, I maintain a life with a concreteness of being. Thus, a cold beer shall continue to retain all of its properties and full potential.

And yet, "*I Have Become Space*" is the title of this book and for the reasoning within as I hope you will come to understand.

Along the way some interesting things may become apparent, as for instance Rate x Time does not equal Distance in an expanding Universe. It only gets us from A to B while B is now C! All this and more will be explained.

Matter has often been described as that which "Occupies space and has mass". The inherent contradiction of this definition is all too apparent unless each is a manifestation of the other. Matter could occupy matter and have no further relationship to space. But as soon as Matter occupies Space, it is by default a manifestation of Space.

CIG Theory is experimentally verifiable. String theory does not appear to be experimentally verifiable. This is a no strings attached unified field theory. It is not a theory of everything. Nothing but GOD is a theory of everything.

CIG Theory's single postulate (which appears to introduce consciousness, a different topic altogether):

Believe It or Not

So, think, read, learn, enjoy, think some more, and take away from your experience a newfound peak into a new view of nature. There will be some overlap in the book, since practice, practice, practice makes perfect. And, on both scales large and small, the topics are all related. Perfect your knowledge and discuss CIG Theory at the next cocktail party.

Go ahead and spill the drink.

End Preface

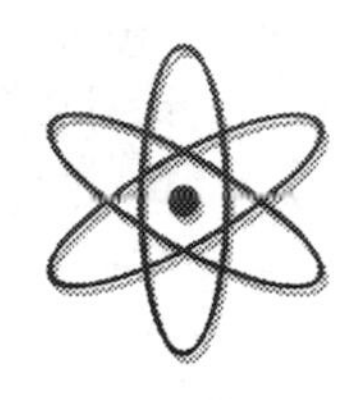

PART I—THE QUANTUM

Quantum Interpretation #1: The Double Slit

The Double Slit Experiment Explained:

With respect to the double slit experiment, if the photon/electron, etc., being in its collapsed state (i.e. black hole-like) prior to departure from its originating aperture, then proceeds to it becoming more spatial (MT=S in accordance with Coney Island Green), manifesting into a much larger three dimensional spatial state, could it then not go through both slits? For instance, without the need for the current quantum conundrum of splitting into two particles and having one particle go through both slits, like some sort of quantum magic? Current thought requires the particle to split and go through both slits, to be in two places at once, like magic. CIG theory allows some of the particle, now in its spatial state, to go through one slit, and some of the particle to go through the remaining slit. No magic necessary. Much more believable.

The introduction of MT=S allows for the particle to go through both slits without the need for quantum confusion. Then, the spatial field faces diffraction, and constructive and destructive interference according to normal wave theory. It collapses accordingly when it "hits the wall", adhering to probabilities of wave function collapse (the probabilities may actually follow the 4th Law of Motion & the desire to reach time equilibrium). The collapse of the spatial field (S/T=M), reintroduces not the particle, but a new particle, only different from the first via its transmutations through its field interactions during its journey from the originating aperture, through both slits, subsequent wave interference, and final collapse.

This is exactly what happens in CIG Theory.

And when you finish this book, it is recommended that you revisit this Double Slit to reaffirm you understanding of CIG. Learn the topic. Then, apply CIG. The experimental outcomes will become clear and none-convoluted. You will have mastered the Double Slit, the granddaddy of all quantum experiments.

Quantum Interpretation #2: Quantum Tunneling

CIG Theory—now an explanation of quantum tunneling (the classical particle becomes spatial enough to overcome the classical barrier). Vibrate quickly enough and turn spatial, then find yourself caught between the decks of the ship on the Philadelphia experiment.

From Wikipedia (regarding quantum tunneling):

Quantum tunneling refers to the quantum mechanical phenomenon where a particle tunnels through a barrier that it classically could not surmount.

Hence, the probability of a given particle's existence on the opposite side of an intervening barrier is non-zero, and such particles will appear—with no indication of physically transiting the barrier—on the 'other' (a semantically difficult word in this instance) side with a frequency proportional to this probability.

End Wikipedia Partial Extract

As regards a solution to quantum tunneling, the following was considered as a pre-thought: Why the space does not escape the balloon (discussed later in this book) but does escape the wire with current (thereby creating the external electromagnetic field). It is offered that the "electromagnetic field" constitutes a form of the Dark Matters and is manifested as new Space in accordance with CIG: My response to this pre-thought:

The new space within the expanded balloon is massively Darker Matter (since the particles are not traveling that terribly fast, though fast and numerous enough to offer up new space) as opposed to the electric field escaping from and surrounding a wire with a current going through it (since the rate of travel is great) which manifests into a much darker dark matter (but perhaps not quite Dark Energy). The electric field has been created as new space (i.e. traveling massive particles and CIG Theory). Apparently, in the balloon topic (discussed later in this book) it appears that the matter cannot cross the boundary nor can certain (relatively slow traveling forms of matter) manifested forms of dark matter. They are slow enough that they are still classically here and solid. On the other hand, the Dark Energy can (pure space) as well as other fast traveling particles that become spatially non-matter enough to cross (tunnel through) the barrier. This pre-thought, right or wrong was the conceptual beginning of my offered solution to Quantum Tunneling, again using CIG Theory, and as follows:

The particle tunnels through a barrier that it classically could not surmount because it travels fast and becomes predominantly space. As space, it crosses the boundary rather readily. It can reappear on the other side. It re-emerges as the classical particle again once it slows to more classical rates of travel. CIG Theory at work. The how of Quantum tunneling was just explained? Yes, CIG explains quantum tunneling.

Quantum Interpretation #3: Bohr Orbitals

The Theory views new Bohr "orbits"/ "energy levels" as creating that space at the expense of mass/energy, thus creating space in discrete Planck-like volumes. It is suggested that someone, using the quantification number (CUPI) provided in CIG, re-calculate the resultant energy level volumetric changes, and confirm consistency between those now known volumes (new radii/energy levels/wavelengths, etc.), and those energy level volumes the quantification would thus provide, as calculated using CUPI. If there is consistency, this may lend credibility to CIG. Note that where utilizing "known new quantum energy levels with their known new radii", to compare with calculated quantum volumetric spatial changes using the CUPI conversion factor (quantification) and known energy levels, please don't forget to compensate for density variations as had been done when formulating the CUPI. [All explained later in this book]

That quantum jumps as interpreted by the MTS equation now take into consideration the discrete spatial changes of Bohr's new orbital configurations, at the expense of mass, simultaneously places physics back into Einstein's world view, namely that: "God does not play dice with the Universe". CIG concepts bring determinism back into the picture by redefining Spooky action at a Distance and discrete Planck jumps as the continuous unfolding of new spatial volumes.

MTS offers that new Bohr orbitals are the creation of new space that wasn't there before.

The photon(s) is the end result of an electron(s) traveling at light speed. The new Bohr orbital is that newly created space. Photons are electrons in motion at light speed! As mass disappears, space unfolds. Attempt to calculate the resultant volumetric space of an unfolding electron, using the CIG quantification, the mass of the electron, and its rate of travel.

Quantum jumps are no longer discrete "here and then there with nothing in between" wow what was Bohr thinking realities. Einstein was right all along. CIG appears to vindicate Einstein. It's no wonder he ignored mainstream science in his later days and worked on the problem until his last days. There was something more to the puzzle. It is truly believed that the Great Professor would be intrigued with CIG theory, perhaps supportive.

The more mass that travels, the more Space that manifests. What CIG is trying to say is that Red Shift anomalies can be explained by recognizing that each stellar entity is essentially its own Big Bang (hence CIG may also be known as the Mini Bang theory), as is each new Bohr orbital. The MTS process works not only at the stellar scales, but at the atomic quantum level as well, and in between. The new Bohr orbitals represent new Space, not simply a repositioning of space.

Red shifts will be explained in the Cosmological section.

Quantum Interpretation #4: Superposition

Wikipedia states "It's well established that the physical aspects of quantum experiments, such as a particle's position or momentum, are not well defined before they are measured."

CIG commentary to the above: that is because they are literally in their spatial state; MTS

Wikipedia further states: "A key idea to achieve such a situation involves the fact that quantum mechanics allows objects to exist in superposition, so that they can be in two or more contradictory states simultaneously; for instance, an electron could be in two different places at the same time."

CIG commentary: The Electron is not, nor is anything else, in two places at once (except perhaps my dear wife); the electron has become spatially manifested in accordance with CIG (explanation of double slit), and the "probability" is real and spatial, only collapsing when the particle is measured (it must be stopped to be measured). Try and capture the particle moving (an impossibility) and you will find a much larger spatial field (thus the perception of superposition).

CIG theory places Superposition back into reality. However, one must first convince oneself that CIG is correct, and just how does matter turn to space?

We will attempt to answer that later. For now, here is the equivalency of one atomic mass unit to its equivalent energy to its equivalent volume of space.

0.02762u = 25.7MeV= 14,952,942.08 pm cubed of space

(Mass) (Energy) (Space)

For the true physicist: Follow Lorentz transformations for rates of travel less than "c"

Quantum Interpretation #5: Decoherence

Regarding multiplicity of states simultaneously:

Think matter to its spatial equivalency here [as matter travels at the rates % "c" of course (MTS and CIG) as it attains its spatial equivalency]. The "multiplicity of states simultaneously" may be viewed according to that "new space" resultant from the traveling mass.

Regarding "The mere act of measurement forces the system to choose a single state of being".

The act of measurement relies on the fact that the "object being measured" remains still, at or near classical rates of travel near zero % "c". This forces the spatially manifested equivalency (when traveling at greater rates) to a single point particle again, one that can be measured. This is the true and real collapse, the decoherence of the "probability" (Per CIG: Not a probability at all but rather a reality of new space) to the measurable point particle. The decoherence is true and real and it is the measurement which must essentially stop the particle in its spatial state that turns the particle "probability" (actually true and not mathematical) back toward its black hole or classical state.

The particle's Quantum state is a new volume of space and not a "probability". The Quantum state decoheres to smaller volumes as it slows and regains mass.

Recall the CIG solution to the Double Slit, wherein the tiny particle becomes spatially large and goes through both slits (a portion of the now spatial electron through one, a portion through the other) without the "probability" and particle being theoretically everywhere and in two places at once as is current thought in physics. CIG takes the fast moving spatially large particle and turns the probability into a reality. This paradigm shift as offered by CIG solves the Double Slit confusion and the decoherence is accompanied by volumetric changes in the particles' size.

The new space offered up in CIG Theory retains information of the original particle however in its spatial form. That retention of information is what allows the particle to re-emerge when it slows down and decoheres back into the original particle, or at least as original as it can be (i.e. a bit here or there may have been dropped along the way). CIG Theory suggests that the spatial equivalency of matter retains the original information, a highly desirable feature. In what form does that information take? Ones, zeroes, curvature bits, time ticks; how are things remembered such that a cohesive decohering collapse takes place, one which retains the original information? The information bites and bits must be found within the MTS equation, as this is where everything resides (Black Hole, Standard Model, Vacuum Energy, etc.). Suggestions anyone?

Quantum Interpretation #6: Non-Locality & Entanglement

CIG Theory is a local theory and does not support the concept of non-locality.

Everything occurs locally, although in CIG that locality is now smeared out over a new and very real volume of Space. The greater the motion, the greater the volume of Space. There may be the appearance of non-locality as the new spatial volumes can overlap with the spatial volumes of other moving particles, and they may "entangle" to a certain extent.

There may even be the appearance that a single particle entangles with itself. But it is only appearance.

In the cases where a particle may be split and half stays where it is and half goes to a location far away and exceeding speed of light communication (i.e. they cannot talk to one another), if spin is found up on the half that stayed home, intuition and not entanglement determines that spin is down at the other location.

If I have a dollar in coins, and split it between two piggy banks, and one piggy bank stays home, and the other sent wee wee far away, they are only entangled to the point we are accustomed to (i.e. obvious sense).

CIG does not subscribe to non-locality and Entanglement. There is no "Spooky Action at a Distance". Here I am in agreement with the Great Professor.

Quantum Interpretation #7: Uncertainty

Let's now pretend to discuss MTS as regards the Heisenberg uncertainty principle (states a fundamental limit on the accuracy of position/momentum as you may know).

But . . . . Now that a moving particle turns spatial (CIG Theory), the concept of exact position per the Uncertainty Principle is no longer natural, nor true. Position is now defined by a much larger spatial volume. It's no wonder we didn't know where the particle (position) went as we defined its momentum to a greater degree. For as its momentum increased (Per CIG as rate goes up mass goes down), it's position (the smeared out probability) actually offered up new spatial volumes. Where is the particle? Literally everywhere the CUPI quantification allows it to be.

The Heisenberg Uncertainty Principle therefore now has a new CIG Interpretation: The particle changes size according to a rate based equation, so it is bigger when faster (and more vacuum like).

Stated another way through the CIG *Certainty Principle*, note the following:

If we know the Rate [% "c" (also represents "T" in the MTS equation)] of the traveling Mass ["M" in the MTS equation], and the mass' starting "black hole" still (non-moving) location, we can know where it is, and how spatial ["S" in the MTS equation] it has become.

Under the Uncertainty Principle, it used to be that the more we knew a particle's momentum ($p=mv$, where m = mass, v = velocity, and p = momentum), the less we knew its position, because . . . . at that time in science, we didn't understand that the particle grew in size, as its mass turned to space.

Quantum Interpretation #8: Quantum Gravity

The concept of Quantum Gravity is applicable to both the Quantum and Cosmological sections of this book and since Quantum Gravity essentially combines the two worlds.

Basically, where Albert Einstein stopped with matter warping the space-time continuum, CIG theory takes the next logical step and offers that it is the space time continuum itself that actually turns into matter. In a nutshell, this is what the MTS equation does.

To avoid duplication, please see Cosmological Interpretation #9 where quantum gravity is more fully discussed.

MTS

Within this equation space, time, and matter, and all the curvatures, from Black Holes to the Standard Model to the Vacuum exist. Within this equation resides gravity in all its curvatures including the strong force (greater curvature), weak force (less curvature than strong force), and electromagnetic force,

The matter is the fabric of space-time.

Quantum Interpretation #9: Conservation of Energy

CIG*:*

1) Offers a New Explanation of Pressure

2) Is Relativistic in nature and therefore builds upon current science

3) May rely on an unfolding of Professor Lisa Randall's extra dimensions to account for the new spatial volumes. One offering attributable to Giacomo, should he wish to take credit, is that the particle offers up new geometric sides to itself, and that this new surface area accounts for the additional space. This makes sense to me. But, the "how" of CIG theory needs to be more fully addressed. [Perhaps the particle travels so fast it is both here, and there, and over there, and here, behind you, in front of you and under the couch seat cushion, such that space is created]. *That* it happens as offered and understood is explained and documented herein to the best of my current ability. How it happens I don't know and don't pretend to know. Hey, look, a nickel.

4) Does not rely on speeds greater than "c" as does current inflationary theory

5) Combines the Fundamentals (Matter, Time, Space)

6) But most importantly, CIG Theory coherently respects conservation of energy whereas the current view of expansion of space does not, as the current view offers up that space and its inherent vacuum energy without taking the energy from elsewhere. Rather, in CIG theory, the new Space comes at a cost, namely mass (M or matter) through motion (T or % "c"). How can current physics stand unless the new space of Hubble's expanding Universe with its associated energy is taken from elsewhere? Current views do not respect the Law of Conservation of Energy. CIG allows for new volumes of energy filled Space (The Expanding Universe) while conserving the Conservation of Energy Law according to the following:

0.02762u	=	25.7MeV	=	14,952,942.08 pm cubed of space
(Mass)	=	(Energy)	=	(Space)

7) The Conservation of Energy topic could have appeared in the Cosmological Section. But . . . alas . . .

Quantum Interpretation #10: Physical Interpretation of $E=mc^2$

The Coney Island Green Theory's Explanation as to Why $E=mc^2$

In conjunction with the reading of this topic, please be advised that to understand Why $E=mc^2$, conceptually speaking that is, one must first have at least some understanding of the nature of light, that is, as its rate "c" is understood to affect TIME (think the old spacecraft trip where your brother goes out into outer space at the speed of light and comes back still as a baby, while here on Earth you're an old man). It will also help if you have read and understand CIG and "Believe It"! If you do not "Believe It", you may not believe in this explanation as to why $E=mc^2$.

So, with the idea in mind that you have some understanding of relativity, here is Why $E=mc^2$.

$E=mc^2$ because all the energy (motion) for a closed loop "there & back", of matter (mass) @ "c" value represents {and here I shall use Big M, to represent all the Matter (mass) in the Universe}, that Matter proceeding to its turning into Space (MT=S) and then that resultant Space turning back again into Matter (S/T=M). The loop has made a full circle, Matter to Space, and back again to Matter, thus representing all the possible motion/movement (Energy) possible in the Universe.

For the trained Physicist/Mathematician: For any given mass of matter, to obtain its spatial equivalency at "c", use the CUPI quantification. For rates less than "c" apply the Lorentz proportions associated with time dilation or length contraction along the x-axis. For known volumes of Space, use the CUPI quantification and work backwards to determine its mass equivalent.

Matter to Space, Space to Matter! We start with mass, proceed through all forward vector TIME, then back through reverse vector TIME, and so all the ENERGY=MATTER x TIME (forward vector) x TIME (reverse vector), or, $E=MT^2$, equivalent to $E=mc^2$, for those understanding time as it relates to "c".

So, all of "E", the one single full closed loop being "Matter to Space and Back Again into that Same Matter" {(MT=S), then M=S/T}is equivalent to $E=MT^2$ (your $E=mc^2$), and conceptually, this is *Why* $E=mc^2$! Remember that "T" is "c", since as one reaches "c" value, one becomes Time itself, in that at "c" perspective, nothing else moves in relation to "c", and "no movement" is equivalent to "nothing has changed" is equivalent to "nothing has happened" is equivalent to "no time has transpired"; only TIME itself doesn't age.

Energy in $E=mc^2$ is replaced in MTS philosophy by the actions of new volumes of space with its subsequent collapse back to matter. When a nuclear explosion occurs, the winds created are the Energy in $E=mc^2$ but in reality it is represented in MTS as new volumes of space that displace the existing environment and the trees and houses blow down, much like the cozy homes of the first two pigs. Recall the third pig made his home of brick. But, as Richard Feynman would say, "I digress", and here I digress a bit farther than most.

$E = mc^2 = MTS$

Energy in the $E = mc^2$ equation is associated with the conversion into new Space in the MTS equation. Mass (m) in the $E = mc^2$ equation is associated with M (mass/Matter) in the MTS equation. And, c^2 in the $E = mc^2$ equation is associated with T in the MTS equation. Recall that T represents forward/reverse vector Time, or zero mph to "c" and back again to zero mph.

All the energy (motion) possible, Matter to Space, back to Matter (all movement possible is all Energy!).

The *why* of $E=mc^2$ has just been explained.

Within the MTS equation are all the available space-time densities possible, essentially the entire Standard Model, and as we shall see in the Cosmological section, Black Holes, Dark Matter and Dark Energy as well.

Quantum Interpretation #11: Collapse of the Wave Function & The Measurement Problem

The introduction of MT=S allows for the particle to go through both slits without the need for quantum confusion. This we have seen in the Double Slit section of this book.

The spatial field faces constructive and destructive interference, according to normal wave theory. It collapses accordingly, adhering to probabilities of wave function collapse (the probabilities actually follow the 4th Law of Motion & the desire to reach time equilibrium—see Annex 1). The collapse of the spatial field (S/T=M), reintroduces not the particle, but a new particle, only different from the first via its transmutations through its field interaction's during its journey from the originating aperture, through the slits, wave interference, and final collapse. The collapse of the wave function is a reality. It is the "S/T" philosophy, reality.

The probability wave function is interpreted as being real (i.e. it is not a mathematical probability but a reality), with the "electron" smeared out (into it being newly created Space), collapsing only when it slows down (i.e. the black hole or "M" side of the MTS equation).

On the quantum level, the discrete "N" jumps no longer can be perceived as simply "here then there with nothing in between activities", as CIG reveals that the jumps actually represent a continuous action (that of the creation of space at the expense of mass). This view appears to put determinism back into quantum. In essence, it takes what is the wave function probability, and lends reality to it, with reason and rationale.

The particle travels so fast it is both here, and there, and over there, and here, such that the wave function probability is an actuality. The Space becomes. Are the particles then, as Giacomo suggests, offering up new geometric sides to themselves that are otherwise not present at slow rates of travel, with those new surface areas (Randall Dimensions) contributing to volumetric space only at higher rates of travel? It appears so. I know of no better explanation. One may say that the "Randall Dimensions" we so often hear about have found a way to unfold. Their collapse back into the "less Spatial, corpuscular particulate form of matter" represents reality when it snails down.

From WIKIPEDIA:

The measurement problem in quantum mechanics is the unresolved problem of how (or if) wavefunction collapse occurs. The inability to observe this process directly has given rise to different interpretations of quantum mechanics, and poses a key set of questions that each interpretation must answer. The wavefunction in quantum mechanics evolves deterministically according to the Schrödinger equation as a linear superposition of different states, but actual measurements always find the physical system in a definite state. Any future evolution is based on the state the system was discovered to be in when the measurement was made, meaning that the measurement "did something" to the process under examination. Whatever that "something" may be does not appear to be explained by the basic theory.

To express matters differently (to paraphrase Steven Weinberg [1][2]), the Schrödinger wave equation determines the wavefunction at any later time. If observers and their measuring apparatus are themselves described by a deterministic wave function, why can we not predict precise results for measurements, but only probabilities? As a general question: How can one establish a correspondence between quantum and classical reality?[3]

END WIKIPEDIA

ENTER CIG

This is how one can establish a correspondence between quantum and classical reality:

As we have seen in CIG, the correspondence between the quantum and the classical is rate dependent. True classical is zero mph and exists only as a Black Hole singularity. Once mass moves, it is no longer classical but is now quantum (though we still view matter as classical until things start moving truly fast). As a quantum spatial field, things can experience interference, both constructive and destructive.

Wherein it is stated "actual measurements always find the physical system in a definite state. Any future evolution is based on the state the system was discovered to be in when the measurement was made, meaning that the measurement "did something" to the process under examination.":

YES, the "did something" was to slow the "real" probabilistic wave function and to collapse it. To measure it you had to slow it, to collapse it. To observe interferes sufficiently and truly so that the probability collapses into the defined observation. In reality there is a new and real spatial volume and not simply a complex mathematical probability. The spatial volume that collapses when measured is customarily seen on the screen in the Double Slit experiment, or even before that if observation posts are placed in between the slits and the screen. Anytime the wave slows, it will become the smaller identifiable particle. Prior to observation, it travels at great speeds and its spatial qualities are manifested. The spatial manifestation of the particle can experience constructive and destructive interference. It has become its larger spatial self. The correspondence between quantum and classical reality relies on the fact that quantum is in motion and classical is zero % "c". Quantum is classical at zero velocity. MTS

The question may arise as to why the particle always finds itself at that exact location at which you measure it, as if by magic you always measure it at just the right location where it is, with no need to search about. This is explained as follows: The spatial particle always collapses at the measurement point since the measurement point introduces the solid stop point, essentially that point analogous to the screen in our Double Slit experiment, the spot that the Mr. Hyde-wave-existence split personality side of the duality requires, demands, insists upon, for it to fully collapse and be seen (the measurement) so as to maintain Time Equilibrium (alternately, motion, pressure, space-time curvature, temperature) with that exact locale. It will always appear where you measure it since you provide the black hole-like stationary field point position to collapse it at that very spot and nowhere else. Where is the particle? Wherever you measure it.

Why does the particle always find itself at the exact location where you measure it?

The following offers a Classification Scheme or Model, [Wiki—Interpretations of Quantum Mechanics], of CIG, and whether or not, the theory falls in line with the offered class:

1) COUNTERFACTUAL DEFINITENESS (CFD) is the ability to speak meaningfully of the definiteness of the results of measurements that have not been performed (i.e. the ability to assume the existence of objects, and properties of objects, even when they have not been measured)

YES = CIG abides = knowledge of the MTS parameters will offer a meaningful picture of reality, properties, etc. even if they have not been measured.

2) LOCAL—The principle of locality states that an object is influenced directly only by its immediate surroundings

YES = CIG is a local theory

In the context of quantum mechanics, superdeterminism is a term that has been used to describe a hypothetical class of theories which evade Bell's theorem by virtue of being completely deterministic. Bell's theorem depends on the assumption of counterfactual definiteness, which technically does not apply to deterministic theories. It is conceivable, but arguably unlikely, that someone could exploit this loophole to construct a local hidden variable theory that reproduces the predictions of quantum mechanics.

YES = CIG is both superdeterministic & is CFD, thereby constructing a local hidden variable theory that predicts QM

It is arguably likely that CIG has constructed a local hidden variable theory that reproduces the predictions of quantum mechanics. The variables are no longer hidden.

3) COLLAPSING WAVEFUNCTIONS—In quantum mechanics, wave function collapse (also called collapse of the state vector or reduction of the wave packet) is the phenomenon in which a wave function-initially in a superposition of several different possible Eigen states—appears to reduce to a single one of those states after interaction with an observer. In simplified terms, it is the reduction of the physical possibilities into a single possibility as seen by an observer.

YES = There is a true collapse of Space

4) HIDDEN VARIABLES (I love a good game of hide and seek)—Historically, in physics, hidden variable theories were espoused by some physicists who argued that the state of a physical system, as formulated by quantum mechanics, does not give a complete description for the system; i.e., that quantum mechanics is ultimately incorrect, and that a correct theory would provide descriptive categories to account for all observable behavior and thus avoid any indeterminism.

NO = With CIG, Nothing needs be hidden—the variable was found (MTS) EINSTEIN vindicated

5) WAVEFUNCTION REAL—A wave function or wavefunction is a probability amplitude in quantum mechanics describing the quantum state of a particle and how it behaves.

YES = wave function real and not a probability, and is rate based

7) DETERMINISTIC—Determinism is a philosophy stating that for everything that happens there are conditions such that, given those conditions, nothing else could happen.

YES = GOD plays dice with Einstein

8) UNIQUE HISTORY—

YES = particle and wave dependent upon % "c" = explains Dark Matter, Dark Energy

9) OBSERVER ROLE—

YES = based on motion = assume a stationary observer

Quantum Interpretation #12: Virtual Particles

Space is not really "empty" but is teeming with virtual particles.

Space can and does have energy of its own.

Virtual particles collapse from their spatial form and become massive particles.

This collapse of space to massive particles and massive particles to space happen routinely and are commonplace.

CIG may explain virtual particles.

Quantum Interpretation #13: The Quantification of Mass to Space

The following is verbatim from the day the quantification first appeared. It is objective evidence that new Bohr orbitals represent new volumes of space and not simply a rearrangement of the existing volumes. It ties in with pressure, and Chemistry, and Dark Energy and more. I have reserved the unit CUPI with NIST to honor my wife.

The math is rudimentary, and is presented more or less as it originally appeared in handwritten notes. The math will remain rudimentary until such time as others articulate in the field may choose to present it in a more eloquent manner, refining the calculations accordingly, and further clarifying the intentions of the author (i.e. author's attempt at a crude countercheck on the astronomical scale, using calculated values of the CUPI, with its defined value as 1u = 541,380,958.7 pm cubed, and a known mass of the Universe).

The Approach:

Using page 1045 Chapter 44 of College Physics (half of which I don't understand yet!) 7th Edition Sears/Zemansky/Young Nuclear Fusion 44-7 Where Four Hydrogen atoms proceed to One Helium atom + loss of mass & release of energy.

1. I equated spatial volumes compensated by density to find the difference in spherical volumes of Helium & Hydrogen.

Then, I equated this spatial difference to a loss of mass & release of energy (25.7 MeV).

This is the Missing Matter of the Universe!! & proves II. Matter x Time = Space (MT=S)

Mass to Space via

0.02762 u = 14,952,942.08 pm cubed

Or, 1u = 541,380,958.7 pm cubed

Mass to space!!!!

The math:

Values From page 1045 Fusion 7th Edition Sears/Zemansky/ Young Nuclear Fusion 44-7

Mass of 4 Hydrogen = 4.03132u

Mass of 1 Helium=4.00370u

Mass difference= 0.02762u

Energy= 25.7 MeV

Atomic Radius (from Googled sites)

Helium= 128pm

Hydrogen= 37.3pm ratio of 3.431 to 1

Density Helium = .1785

Hydrogen = .08988

= ratio of 1.98598 to 1

Note: make it easy & use two to one ratio

Calculate Spherical volume 4 pi r cubed/3

Helium = (4) (3.14) (128 cubed) /3

Hydrogen (4) (3.14) (37.3 cubed) /3

Helium = 8,780,076.373 pm cubed

Hydrogen = 651,802.67 pm cubed

Then: multiply by four hydrogen atoms (four were in the fusion process)

= 2,607,210.67 pm cubed

Atomic spherical spatial volume of four H & one He without density compensation.

H-	2,607,210.67	pm cubed	pico meters cubed
He-	8,780,076.373	pm cubed	

Then: Compensate for Density

Density Helium = .1785

Hydrogen = .08988

Since Helium is denser, to compensate, we find average spatial density (2 to 1) & double the Helium volume so that—
8,780,076.373 pm cubed x 2 = 17,560,152.75 pm cubed

Subtract the volume of Hydrogen!!

17,560,152.75 pm cubed minus 2,607,210.67 pm cubed = 14,952,942.08 pm cubed

Ergo:

Mass 0.02762u has turned into 14,952,942.08 pm cubed of Space!!

Equating energy to mass to space

0.02762u = 25.7MeV= 14,952,942.08 pm cubed of space!!!!!!

(Mass) (Energy) (Space)

Internal Notes: same molar volume for H & He of 22423.5 cm cubed per mole

1u = unified atomic mass or Dalton (Da)

1u = 1.66053 x 10 to the negative 27 Kg = about 931.49 MeV

Mass of known Universe 1.6 x 10 60th kg

Volume of Universe 4.2 x 10 69 cubic miles (these values googled)

1u = 541,380,958.7 pm cubed

using mass above 1.6 x10 60th/ 10 to the minus 27th = 1.6 x 10 87th

1.6 x 10 87th x 1.4 x10 7th = universal volume of 2.24 x 10 94th pm cubed

What we do with the gamma radiation is described in *The Theory* and relates to the actual mechanism by which space is created from matter, and explains dark matter/dark energy.

In any event, through the use of atomic radius, nuclear fusion values, page 1045, "Googled" values of radius & density, calculated spherical volumes, finding difference, compensating for density, We Have;

Statistical Confirmation that

1u = 541,380,958.7 pm cubed
*   *   *MASS to SPACE*   *   *

Ergo : the continuum broken!! Breaking the space-time continuum is key to *The Theory.*

Internal Note:

The Space-Time continuum must be broken for matter to become space.

Documented March 25, 2007

Discovered Saturday night March 24, 2007 at about 10 pm EST

Finalized by 1 am Sunday

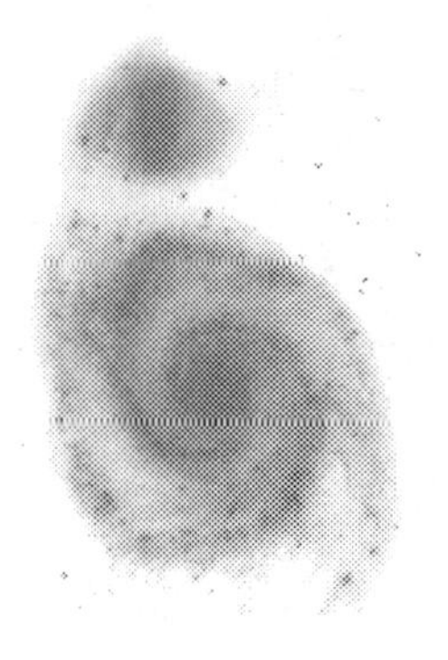

PART II—THE COSMOLOGICAL

Cosmological Interpretation #1: Dark Matter & Dark Energy

It will make sense to learn a bit about Dark Matter and Dark Energy before approaching this topic. A quick online GOOGLE search will offer much. Dark Matter and Dark Energy are placed under one topic roof as CIG teaches that they are simply forms of one another, the same matter at different rates of travel, manifesting into their different spatial quantities but with their own quantum, cosmological, and gravitational qualities (curvatures, temperatures, etc.).

In short, Dark Matter is the gravitationally pulling matter that prevents the stellar stars n' swirls surrounding galactic cores from flying off into outer space, as the observable matter is insufficient to keep them gravitationally inbound. Based on how fast they are found to be revolving around the galactic core the stars should be flying off. But, they are not. So, Dark Matter accounts for the needed matter to keep the galaxy stable. Dark Energy is the energy associated with the expansion of the Universe. Regular matter accounts for only about 4.9% of all energy, while Dark Matter 26.8% and Dark Energy 68.3% (percentages retrieved from Wikipedia). The Cosmologists are the experts here.

Per CIG, Dark Matter and Dark Energy are simply the same observations but viewed from different perspectives of % of "c" travel. The halos surrounding massive bodies are examples where due to gravitational effects, these effects limit the "c" rate sufficiently such that a different "Spatial" quantity with a different "mass component" unfolds. The space unfolds in the same proportions as length contraction along the x-axis.

One can look at the dark energy/dark matter data and attempt a confirmation of the correlation between Lorentz Transformations, Time Dilation and Length Contraction along the x-axis and find that it matches the percentages of unfolding energy (Matter/Dark Matter/Dark Energy) based on the rate of travel of massive particles. When light can't move at all (black hole), no space manifests and mass alone exists in a singularity. This is the true classical Newtonian world.

NOTE: There are other considerations here such as there must be a certain volume of space for mass to exist; conversely, for the vacuum energy, there must be a mass component, since singularities appear unstable.

a) On Dark Matter/Dark Energy Specifically

According to "*The Theory*" (CIG), in line with length contraction, relativity, Lorentz transformations, Mass to Space conversion, etc. the dark matter (i.e. 26.8% of energy) is created by matter approaching but not reaching the speed of light, whereas the dark energy (i.e. 68.3% of energy) is created by matter reaching the last percentages (i.e. 90-100%) of the speed of light. At the speed of light, one becomes time (this, since nothing moves relative to this going rate speed "c", so all stands still relatively speaking, and stillness means that no Time has transpired, much in the same way that a house left by its occupants in the year 1910 will have preserved its time within that era). So the equation MT=S reduces to matter at the speed of light [(i.e. multiplied by Time) (MT in the equation represents matter at the speed of light—coincident with Rate x Time = Distance, so MT is photon at the rate of light (mass @ "c") = Space)] is equal to Space. Matter to Space! Theologically speaking, "God is light". The God-given light gives us space, time & matter—to please the Vatican.

Hence the dark matter (the gravitational pulling matter) is found at slower speeds whereas the dark energy (the expanding matter responsible for the expansion of the universe—Hubble et. al) is a direct result of the matter at much higher speeds (those approaching the speed of light). Essentially, there is no definite boundary between the two (dark matter & dark energy), and one must think of dense matter at zero rest speeds (i.e. you & me), with spatial matter (space itself) reached at the speed of light. In between, there are essentially many (perhaps infinite?) of these "forms of matter", dependent upon the rate of speed of the matter. To light, all stands still and there is no Time, for if there is no movement, nothing has changed by which time can be measured. Time is of course movement (rate) dependent.

With regard to the Lorentz association with Dark Matter to Dark Energy: Matter unfolds into Space—and depending upon matter reaching what exact percentage of "c", it either becomes more massive space (Dark Matter = lower % "c" rates), or more spatial matter space (Dark Energy w/ expansion properties), with pure space unfolding at the rate "c"!, this since as one reaches "c", one becomes "Time", and as viewed by the citizen at rest (you & I), matter becomes space (MT=S). Duration (i.e. mm:ss) is not the time under discussion. In our theory, time only takes place if/when there is motion. Of course, all events take place over a duration.

If there was a primordial near infinite point mass of matter, and it represented all that is, with anything outside the point being nonexistent—neither Time nor Space nor more Matter, and if that point now expands, Space has been created—however, it is only created at the expense of the matter point—mass has been lost, and through Time, Space has been created (again MT=S). Conceptually, this is the Big Bang.

For your enjoyment, this is the theory. Later in this book we shall offer the "fourth law of motion". Right now, we have quantified mass to space, which explains the missing Dark Matter/Dark Energy and also offers a satisfactory reason as to why the Universe is expanding. Look deep into the explanation/wording of the "fourth law", and one can see that everything can be considered the same entity, the same particle-wave-field-space-time-matter, which simplifies the current confusion.

Of course there is the Final Law of Everything "A" (FLEA) that will always be a work in progress and noted herein conceptually only as an introduction. The Final Law of Everything "A" cannot be finalized even when philosophically categorizing everything as one point, one particle, one consciousness, synthesizing Philosophy, Religion, and Physics et. al . . . That it cannot be finalized is as close to the final law as we shall endeavor to comprehend. Thus the reader must understand now that it is not the intention of this paper to address FLEA any farther than we have already. But I think we're getting ahead of ourselves, don't you?

From the Wikipedia site “Cosmological Constant” the following was excerpted:

“One possible explanation for the small but non-zero value was noted by Steven Weinberg in 1987 following the anthropic principle.[8] Weinberg explains that if the vacuum energy took different values in different domains of the universe, then observers would necessarily measure values similar to that which is observed: the formation of life-supporting structures would be suppressed in domains where the vacuum energy is much larger.”

Specifically, regarding, “Weinberg explains that if the vacuum energy took different values in different domains of the universe”, by now you should recognize that Coney Island Green Theory does exactly that.

It is inherently implied in CIG the ability to explain exactly how the vacuum energy takes on different values in different domains of the universe. Based on varying rates of massive particles of course. The same explanation I am providing for Dark Matter and Dark Energy.

Dark Matter and Dark Energy are different Space-time curvatures. And so is the Standard Model. And each and every particle, its own curvature. Everything is the same, electron, proton, muon and gluon, if only they were traveling at the same rate and thus became identical curvatures.

Dark Matter is one curvature; Dark Energy another. Within the MTS equation exists all curvatures.

Cosmological Interpretation #2: The Horizon Problem

Apparently, to explain the horizon problem, current inflationary theory relies on "freezing" in the constituents at some 10 to the minus 32 seconds, then, during a much "faster than light" inflationary expansion, this "frozen" homogeneity maintained its exact qualities throughout the expansion. Our picture is different, though respects all views of nature.

Enter Coney Island Green Theory (CIG Theory). CIG states that wherever you have mass moving at or near "c" rates, that mass manifests itself into a certain spatial quality with definite temperature characteristics. At say 98.2345 % of "c", regardless of where in the universe that matter is moving, it will have the same space-time curvature, spatial quality, and thus the same temperature. Since the temperature is inherent to the spatial characteristics, one will find the same temperature throughout the universe, where one observes the same mass at rate X. There is no need for opposite sides of the universe to talk to one another nor is there a great need for "inflationary theory" to "freeze" in the qualities long before the universe took its current form.

As I understand it, the homogeneity (temperature/mass distribution/energy, etc.) on both horizons of the Universe is exactly nearly identical. This would occur if the Universe had enough time to "mix"—like warm water poured into a glass of cold water, and let to sit for a day. But, from a single point Big Bang, since light energy emanates to both horizons, and these horizons are found to be the same temperature, the problem is "how can this happen if nothing travels faster than light—meaning, how could the two horizons have time to mix to a homogenous state if there is insufficient time for them to do so", as light has traveled in opposite directions out to each respective horizon (i.e. it would always take twice as long to mix as the time in the universe allowed). This is the "problem".

CIG Theory, as you are aware, states that anywhere mass is traveling at a given rate, it is turning into a spatial quantity with its inherent energy / mass distribution / temperature / space-time curvature associated with it (Rate = Time = Temperature). Therefore, a mass on one side of the horizon, traveling at the same rate as a mass on the opposite horizon, will have identical temperatures, since they have unfolded into the same density space (i.e. Dark Matter or Dark Energy). They don't have to talk to one another to be at the same temp.

Solution to the Horizon Problem Revisited:

Regarding the Horizon Problem: http://en.Wikipediapedia.org/Wikipedia/Horizon_problem

If the very nature of light (matter), per MT =S, always manifests itself according to its intrinsic values [To variations of Space (dark matter, dark energy, etc.), based on rate (c or percentages thereof)—and according to CIG Theory, of course], then, even though the distinct opposite ends of the Universe cannot logically correspond (talk) to one another due to their distances exceeding the limits of time, there shall still be an exact match of their temperature.

In fact, CIG Theory demands this. CIG Theory offers a reason why the universe is in fact extremely homogeneous. It is based on the very nature of the manifestation of mass to space. At pure space (the darkest of dark energy), temperatures are identical, and there is no need for communication between two horizons. Each separate side of the Universe is following the same CIG transformation (mathematically no different from the other when mass travels at an identical rate), so of course they will be at the same temperature.

The Horizon Problem states, in part "This presents a serious problem; if the universe had started with even slightly different temperatures in different areas, then there would simply be no way it could have evened itself out to a common temperature by this point in time." The temperatures are a direct result of the mass to space transformation process. There is no need to "even" anything out. It is inherent to the CIG process.

CIG Theory requires it be, because at the rate "c", any entity will manifest itself into the identical Space, no matter what side of the Universe that entity is on. Therefore, they will be at the exact same temperature. This is the nature of CIG Theory. The same Dark Matter on one horizon will be at the exact same temperature as the same Dark Matter on the opposite Horizon, as its calculations (transformations) are one and the same.

CIG Theory allows for a coherent explanation of the Horizon Problem. In fact, to the CIG Theory, it is no problem at all, since CIG would expect that the universe be formed with precisely the same properties everywhere (space for space, dark matter for dark matter, i.e.—the same transformation is only comparable to its like transformation—you cannot compare the temperature of dark matter with dark energy—here the dark matter will be hotter)(by the very nature of the manifestation itself—the transformation of matter to space). It could also have formed decidedly slower, without an inflationary aspect having to "freeze" in these properties. The properties are inherent to the CIG transformation.

An analogy:

If you bake some bread on one side of the Universe, and you bake some bread on the other side of the Universe, and they taste the same, did the loaves have to talk to each other? CIG Theory says no, because in CIG, we are baking the bread. Current theory says how can the bread taste the same, they must be talking to one another (but they can't, because they exceed the limits of time). Current Theory forgot to bake the bread!

These are just a few thoughts on the Horizon Problem. Please make every effort to understand CIG Theory in this regard.

Remember:

Where there is a different time there must be a different place. Where there is a different place, there is a different space. Where there are different spaces, there are different volumes. CIG theory explains the creation of new volumes of space created as the result of different times imparted onto the world universe and as a direct result of the relativistic nature of nature arising from different rates of traveling massive particles. These rates are limited by "c", and are often simply a percentage of "c". The Michelson-Morley, Lorentz, Einstein work should appear apparent in CIG, at least to those that have a Wikipedia knowledge of them. It also appears justifiable that Bohr orbitals, Planck, and quantum theory have common ground within CIG, especially where the newly formed discrete orbits that are created can be considered new space, in accordance with MTS. The accumulated new space adds up to an expanding Universe. The individual stellar expansions explain the Red Shift (historically, it was the Red Shift data that preceded the concept of expansion—then, following the expansion in reverse, to the concept of the Big Bang). Keeping all of the above in mind, we shall now discuss Red Shift.

Cosmological Interpretation #3: Red Shift & Red Shift Anomalies

This topic requires some small knowledge of the Doppler Effect, and after a brief review of Red Shift, apply that concept here.

MTS is the explanation of where the space comes from, for there to be expansion in the Universe. It all happens on a local level (micro) and this explains Red Shift anomalies as well. It is very important to recognize that the macro world must always be built up from the micro. This element is consistent in CIG. Big things are built upon small things.

A noteworthy comment: *The Theory* (CIG) does build upon the past efforts of others, recognizing and explaining both Hubble's red shift and Einstein's concepts of the curvature of space-time. It is important that a theory has some foundation in existing physics, as present knowledge builds upon the discoveries of others. I should like to think that Einstein would have come to the conclusion of CIG if in 05' he was aware of Hubble's Universal Red Shift expansion. But by the time Hubble's concept of an expanding universe came along, Einstein only thought to consider his Cosmological constant a blunder. In fact, as documented in the Metamorphosis toward the end of this book, CIG's MTS equation is interpreted to be found within Einstein's field equation. It was there all along, only awaiting to be interpreted as in this book.

The expansion explains the Red Shift (historically, it was the Red Shift data that preceded the concept of expansion—then, following the expansion in reverse, to the concept of the Big Bang).

Time dilation is interpreted (a true and real nature) as new space [this is where the expanding Universe comes from] Answer to Red Shift anomalies as well.

The concept of the Big Bang, as CIG employs and extends it, is applicable to each stellar mass (star and/or galaxy cluster, etc.) as well as each atom. One can easily recognize that there is little to distinguish the three (Big Bang, our own sun, any given atom). All three offer up their own space and time that prior to CIG, has only been applied to the Big Bang itself] [the parameter property that offers up the space and time (space-time) is motion. The interplay between mass (matter) and space-time is the percentage of "c" travel. MTS

CIG quote of the day: "If it looks like a BIG BANG, and sounds and smells like a BIG BANG it is a BIG BANG, no matter how small it is"

The matter becomes the space. The curvature becomes the matter. Matter does not warp space-time, it is space-time.

The fabric of Space-time does not tell matter how to move, it is the matter.

The below exists in the form of red shift anomalies:

For the Astrophysicists, *The Theory* offers a reason for the expanding universe as well as a solution to the dark matter problem. One can distinguish the rate of spatial expansion near a large star versus a small one (of lesser mass to energy), and this constitutes physical proof of *The Theory*. Things would be expanding quicker around the large star.

For any astrophysicists looking for stars to compare, wherein it is stated above that there should be more expansion around the large stars, obviously, since all stars are expanding to create space, {wherein it is stated "all stars are expanding to create space" naturally, this is taken to mean "mass depletion to space creation in accordance with MT=S and the CUPI" and not simply to be taken as "all stars are expanding to create space" as in maybe an exploding supernova} the stars (galaxies, systems, etc.) to be compared must be very near each other such that they both see the exact expansion that the other may see by virtue of systems nearby (because if they are in different parts of a system, they would be discombobulated by different expansions from other systems expanding in different ways) Thus, the two stars to be compared must be close enough to be able to state, "we stars see everywhere the exact expansion the other does", thus any difference in their measured expansion rates are attributable to their own spatial expansion rates. The red shift on the larger star should be greater, because it is creating more space than the little star. It is losing more mass/energy; it is moving farther away faster and so the Red Shift will be greater. However, note too the two sources to be compared must not gravitationally disrupt each other such that this interferes with what we are seeking to verify, namely the difference in expansion.

Red Shift. Red Shift anomalies. Doppler Effect. Study. Study. Study.

Yawn . . .

Cosmological Interpretation #4: The Accelerating Expansion of the Universe

The reason the "Universal Expansion" is accelerating is because the expansion (new space) emanates from each individual stellar massive body. From our perspective, the first line of galaxies closest to our own offers the first spatial expansion we see. Add to that the next outer galaxy (adding new space), and the next (again adding more space), and the next (and more), etc. The additive effect of all this new space is the "acceleration" of the expansion we see. The farther out we look, the faster this acceleration will be, and of course the greater the Red Shift. As an analogy, take 500 people in a room each blowing up a balloon. Take the person in the center of the room. He/she sees the expanding universe "accelerating" (all the balloons being filled at once).

The "raisins" are adding to the space. [Raisin loaf analogy we often here on TV]. They are not static entities while the balloon expands. They are in fact the cause of the expansion and at the expense of mass. Mass goes down as its rate of travel increases.

0.02762u	=	25.7MeV	=	14,952,942.08 pico meters cubed of space
(Mass)	=	(Energy)	=	(Space)

Don't forget that for more than one entity, (think bubbles upon bubbles) receding from one location (say EARTH), and each other, we have to add these two (or more) distances together to obtain the accumulated distance (reason for accelerating Universe).

Topic of Google study: Standard Candles and Type 1A Supernovae

MTS at work = Accelerating Universe

Bubbles upon bubbles

Many Balloons

Cosmological Interpretation #5: Varying Cosmological Constant

The Cosmological Constant was Einstein's "greatest blunder" and was added to his field equations to counter Gravity. Since Hubble and the idea of an expanding universe was not known to Albert, and since he believed in a static Universe, the constant was added to exactly counter the pull of gravity.

CIG views the constant as a non-constant.

The cosmological non-constant varies with rate of travel as we have discussed.

With each different rate of travel comes a different space-time curvature, pressure and temperature, and Cosmological non-constant.

It is not a created term as when first created but exists as a placeholder of space-time curvature. For each different curvature there exists a different cosmological non-constant.

Within the MTS equation are all the Cosmological non-constants between a black hole and the vacuum energy.

Cosmological Interpretation #6: Combination of Space-time and the Mass-Energy Equation

The Coney Island Green Theory's Space-Time/ Mass to Energy Equation Genetic Similarities:

Recall that "The Theory" stated that it combined the space-time continuum with the mass-energy equation, to get the resultant MTS equation. The following is some straightforward corroborating evidence.

As regards the space-time continuum and current thought, and the mass energy equation ($E=mc^2$), and the notion that matter warps space-time, it can be seen that since the "Time" portion of the space-time continuum is "c" rate dependent (different velocities give different time relative to stationary observers) and since $E=mc^2$, it is readily apparent that both entities (space-time & $E=mc^2$) correlate/manifest to one another through their common benefactor "c". Since rates vary (velocities at which matter travels), Space-time is not constant but varies according to varying rates of "c" (of course "c" is "c" and I use varying rates of "c" to mean that mass does not always travel at "c". So, consider it a % of "c", that's what we're saying. Space-time is not constant but varies according to varying rates of Time (based on the varying rates of matter, that at pure "c" turns to pure space (the darkest of Dark Energy) while Dark Matter is found at lower "c" values—OK we'll call them "d", "e" and "g" why not. Time is apparent in $E=mc^2$, because Time and "c" are one and the same. Time here is separate from duration, which is different.

Dismantle and analyze the MTS equation and we have TS (alternately ST) or "space-time". And we have MT, or alternately mc^2 and for the following reason: We know that in our equation it is defined that "T" represents forward and reverse vector percentages of "c"; zero to c mph and c to zero mph; this is c^2 (T in the equation). And we also know that M is matter or mass (m). So, mc^2 is also found within the equation as MT. In the manner just explained, Space-time and the mass-energy equation have been combined. This combination makes CIG a theory of Quantum Gravity.

In case you are wondering what happened to the "E" of $E=mc^2$:

As regards Energy or "E", the following is offered:

One can view energy only as an ability to do work (in Joules), however, if mass is used up, gone, poof, one still needs a reaffirmation that the basics are taken care of, namely, time-space-matter, (seconds, meters, Kilograms) for everything shall be forever bound to these three fundamentals. The concept of energy is great ($E=mc^2$), but what about the fundamentals? *The Theory* quantifies mass into a volume of space and makes one feel much more comfortable now that the fundamentals are maintained.

The Energy is of course still there and is in fact conserved. But the Energy is now offered up as the new space. It is just this creation of space and new volumes of space associated with it that are the cause for the explosion in nuclear bombs and other explosive devices. So the E in $E=mc^2$ is still there, though now offered in the MTS equation as new space (i.e. the explosion).

Based on the above, CIG has combined the space-time continuum with the mass-energy equation, to get the resultant MTS equation, and this is Quantum Gravity.

Cosmological Interpretation #7: Black Holes

CIG Commentary: Time ticking is rate dependent. Matter turns to space; space turns to matter; In between is Dark Matter and Dark Energy; Standard Model, etc. etc. All explained rather simply. Light cannot escape a black hole as it has turned into the black hole (discount the horizon radiation) and therefore exists at zero mph. Its "ST" components have become the black hole. This occurs at the zero % "c" or pure "M" (matter) side of the MTS equation. The Black Hole side.

CIG is a quantum gravity theory. It takes gravity and combines that curvature with the weak force, strong force, and electromagnetic force. All curvatures and everything takes place within the MTS equation. In fact, you and I take place within the MTS equation.

Pure gravity = black hole = zero rate = M side of equation = full curvature

(space-time has turned into the black hole)

Vacuum energy = pure space = "c" travel = S side of equation = no curvature

[Matter (black hole) (full curvature) has turned into open Vacuum space-time with little curvature]

The matter is the space-time.

The equation: MTS

Transfers mass to spatial volume.

In accordance with: 1u (atomic mass unit) = 541,380,958.7 pico meters cubed of space

Offers the full spectrum of matter/space: black hole to space

As stated, in between is dark matter, standard model, dark energy, etc. etc.

Works on the very small quantum scale, and the galactic as well.

T = time = % "c"

M = mass (matter)

S = Space (vacuum energy)

MTS (Cig Equation and where T is forward /reverse vector time).

And, coming from relativity, where time slows at or near light speed and its elevator experimental thought equivalence, black hole gravitational fields (at or near), herein exists the balance, the "why" this is so:

Time slows since in its forward vector (high speed) manifestation, it is used up (i.e. there is no faster rate and any given piece of matter cannot travel any more [it has all turned to space]) So, in effect time slows (none left).

Similarly, reverse vector time (black hole & the M side of the equation), mass (or its equivalent space) cannot go slower, it has stopped and all reverse vector time has been used up (all gone!). It is again all matter (space to matter).

So, within the MTS equation, viewing T: on the M side it is used up, on the S side it is used up (pure matter, pure space at the extremes). This represents the physical reality of the relativity equivalence of why one views the symmetrical aspect of Einstein's equivalency of gravity to acceleration. It's all there in the MTS equation.

The above is why the stated equivalency exists as a reality.

The Three Equations:

1. S=MT (Big Bang / New Bohr Orbitals) (forward "T" vector)
2. M=S/T (Big Crunch / Black Holes) (reverse "T" vector)
3. T=S/M (The Means, "T" as an arrow vector quantity)

Everything must be interchangeable. Everything is the same. There is only one particle.

When there is no movement, this is a black hole.

This ends the Black Hole topic. We digressed but never to the point of this little piggy went to market. This little piggy stayed home, nearly on topic.

Cosmological Interpretation #8: The MTS Equation

M = Matter (mass)

T = Forward/reverse vector Time (% "c")

S = Space (the vacuum)

Full gravity = black hole (one time vector direction in MTS)

Full Acceleration = Vacuum Energy (the other time vector in MTS)

MTS = Different Gravitational Fields = Different Pressures = Different Space-time curvatures = Varying Cosmological Constants (non-constants) = Different Spatial Temperatures (think Horizon Problem) = Different Times (dilations) = Different Rates of Travel (% "c" of traveling mass) = Different degrees of Matter (Black Holes, Standard Model, Stellar Masses, Heavy Dark matter, Light Dark matter, Dark Energy, Pure Vacuum Energy) = MTS = Different Volumes of Space (think Red Shift and the expanding/accelerating Universe) = Different Densities of Matter (Black Holes being one) = Different Masses (> rate = less mass) [E = mc^2; hold Energy constant, rate goes up, and mass goes down) = CIG Theory = MTS = Conservation of Energy = Ideal Gas Laws = Einstein's Field Equations = MTS

To have “c”, we must start at zero velocity; for “c” to be massless, it must have started with mass; It lost that mass as it approached rate “c”. The Matter (mass) offered itself up as new Spatial Volumes. MTS.

The Space-time Curvature becomes the Matter. It is one and the same, only different in its degree of being one another.

A sidetrack:

Specific to De Broglie:

Let “X” = wavelength (for ease of symbol)

IF: X=h/p (De Broglie equation), where X = wavelength, h = Planck constant, p = momentum

THEN:

X = h/mv

Because p (momentum) = mass multiplied by velocity {m = mass, v = velocity}

And, (please confirm my math here)

Equivalent equation: m v X = h

Equivalent equation: m = h / vX

Then, what we see is that when “v” gets very large, “m” gets very small;

Where “h” is a constant, and assuming arguendo that wavelength X remains constant,

[Although speed of light “c” does equal wavelength multiplied by frequency]

So, at faster rates of travel, mass gets smaller: m = h / vX [as “v” in the denominator gets bigger, “m” gets smaller]

At “c”, mass disappears altogether [it has turned to Space (MTS & CIG Theory]

We are told by the Great Professor that mass gets infinitely greater toward it traveling at “c”

Here, CIG must respectfully disagree.

Cosmological Interpretation #9: Quantum Gravity

CIG is a quantum gravity theory (a theory that combines quantum mechanics with gravity) as it equates the two:

The space-time turns into the quantum/classical matter. All through % "c"

The matter and the space-time are manifestations of one another.

The fabric of space-time is the matter.

The two can only be distinguished by their density. Their density is based on the mass' rate of travel.

Only at full curvature (zero mph) does quantum become classical. Any rate of travel, and the spatial manifestation can reflect itself as a wave, and offer diffraction properties, constructive and destructive interference. However, at slow rates of travel the spatial manifestation is so insignificant that we are accustomed to considering our immediate existence as classical.

The gravity (in the form of curvature) is the quantum, and the greater the curvature, the less quantum it becomes. Only at the full curvature of the singularity does quantum become classical.

Take apart the MTS equation and we have matter (M) and space (S) and time (T).

Inherent to the space-time curvature is the gravitational field. The various field's curvatures dictate what type of matter exists. Quantum Mechanics exists within MTS as does Gravity in its various strengths. Quantum as M, Space-time as variations on ST.

Within the single equation, 1) *Quantum* exists as matter and 2) *Gravity* exists in its various strengths, and the transition between the two has been stated over and over again in this book. Matter turns to Space, with all the possible curvatures offered up along the way, and where the curvatures are the various forms of matter. CIG Theory marries the two and they cannot be divided.

Do you Quantum take thee Gravity to be your lawful wedded wife, to love and to cherish, to honor and obey . . .

I now pronounce you . . . .

Quantum Gravity

Cosmological Interpretation #10: Sonoluminesence

On Sonoluminescence (about which I know very little and offer the following only as a guess):

The implosion of the bubble, which thus creates the photon release should be consistent with the CUPI quantification.

One can repeat the experiments to show that the larger bubbles create more light, and that their implosion characteristics create that amount of energy consistent with the volumetric properties associated with space to energy equivalencies as offered by CUPI.

And the largest bubble (and the reason this topic appears here and not in the Quantum Section) is the Universe itself. And should the Space Bubble be the Universe itself one day, it can according to CIG implode back into Matter.

Cosmological Interpretation #11: The Balloon

The Coney Island Green Theory's Balloon Experiment

Heat causes things to expand (get bigger) and cold causes things to contract (get smaller). Things such as air will contract and take up less volume when cooled. Similarly, things will expand when they get hot.

Heat causes expansion because it increases the vibrations of a material's atoms or molecules. In a gas, heat also increases the speed at which the atoms or molecules move about. The current philosophy is that the increased movement forces the atoms or molecules farther apart and the body becomes larger. By now, after reading this book and understanding CIG, you should recognize the following:

The Coney Island Green Theory of "The Balloon"

OK, so we have one balloon, tied off so no air can escape, nor get in.

Now picture that same balloon in two different scenarios:

1) one cold deflated and limp, and 2) the second scenario: heat (energy) external to the balloon has been added, making this second scenario filled with an increased volume, expanded because of the heat. These two scenarios are most familiar.

Question? Where did the increase in space come from? (i.e. the new expanded volume in the heated balloon).

We apply external energy to a volume that starts off with a set volume of space, and get a new greater volume of "expanded space". Simple physics? This is currently explained by the scientific community as follows:

Heat causes expansion because it increases the vibrations of a material's atoms or molecules. In a gas, heat also increases the speed at which the atoms or molecules move about. The increased movement forces the atoms or molecules farther apart and the body becomes larger.

However, CIG offers the following:

If we start off with a set volume of space, how could simple particle movement lead to an increase of spatial volume? Are we to believe that particles bouncing around at greater speeds bounce off the interior balloon wall and make it bigger, and that this is where the internal "new internal expanded space" comes from?

And here is the link to CIG as you should now be aware of:

I am more apt to believe that the added energy increases the atomic radii of each of the internal particles in accordance with MTS, and that this creates additional space, with the loss of the external energy, and that the discrete amount of energy added, correlates to increased atomic radii (i.e. space is created!). Through MTS we understand the reason why there is more spatial volume inside the balloon. Energy to Space.

Current science wants us to believe that the particles are simply bouncing off the walls more, and into each other, forcing them farther apart. Farther apart from what we (now that you know CIG Theory) ask??!!—we started off with a set volume of space—how does the concept of particles simply bouncing off each other magically create more space??!! How can this create additional space inside the balloon??!! (It can't).

Should there be any doubt:

We have to put forward the accepted definition of "Space". We can live with the Wikipedia partial definition as follows: "Space is the boundless, three-dimensional extent in which objects and events occur and have relative position and direction", though we may believe it has bounds beyond which even it does not exist, nor anything, but this is another topic altogether.

Therefore, when we see an enclosed system, such as a cold balloon, with a defined region (internal volume) of space, and when that balloon is heated, we see a new internal volume greater than that which existed before we started heating the balloon. We are told by physicists that the molecules are moving faster, farther away from one another, and bouncing off the walls of the balloon thereby expanding the volume (at the expense of density of course).

Our problem with this, is that when maintaining the definition of Space we started with, we need an explanation. We cannot fathom where the new "Space" came from without introducing a reason. We continue to respect the definitions we have used when entering into the "experiment". Where did the new space come from?

To keep it short, we have new "spatial volumes", do we not? The balloon is larger, yes? And it is an enclosed entity, yes. It is our experiment. The only thing that happened is that the particles are now moving more rapidly. Now that we know CIG, we associate this movement to the "creation" of this new "space". In maintaining the Conservation of Energy Laws, something must be lost for the new "Space" to arrive. This we attribute to a loss of matter and in a similar fashion wherein mass is lost when energy is created in the $E=mc^2$ equation.

It is also the explanation of where the space comes from, for there to be expansion in the Universe. It all happens on a local level (micro) and this explains Red Shift anomalies as well. It is very important to recognize that the macro world must always be built up from the micro. This element is consistent with CIG.

To explain the Double Slit, we approach the experiment from the view of CIG, and we now have no problem understanding the outcome(s). Light is a "particle" when stopped and in its "black hole" like personality (i.e. at the point of emission, and at the point where it again is absorbed by the screen). However, when traveling, its mass becomes spatial and part of it goes through each slit, thereby offering the opportunity for constructive and destructive interference. [There should be a point where if the slits are spaced far enough apart, that the spatial manifestation, since it is limited by the mass to space conversion limitations/

potential (see CUPI quantification), will be unable to go through both slits and interference will not be possible, since only a limited amount of spatial volume was created and it was not large enough to span distant slits.].

We know this, and we know why balloons expand in a hot environment, because we know and believe in CIG.

As for Dark Matter and Dark Energy, they are simply the same observations but viewed from different perspectives of % of "c" travel. The halos surrounding massive bodies are examples where due to gravitational effects, these effects limit the "c" rate sufficiently such that a different "Spatial" quantity with a different "mass component" unfolds. The space unfolds in the same proportions as length contraction along the x-axis based on how fast a massive particle travels.

When light can't move at all (black hole), no space manifests and mass alone exists in a singularity. [there are other considerations here such as there must be a certain volume of space for mass to exist; conversely, for the vacuum energy, there must be a mass component—singularities appear unstable—but we will not get into that here, again].

We now walk the streets knowing we have a certain equivalency to space and time. It should be an uneasy feeling. We now know CIG combines the space-time continuum with the mass-energy equation. We now look up at the stars and recognize that the space around it unfolds from the star itself.

Space is emergent, and manifests itself from traveling massive particles. Look inside the balloon, not outside. There is more going on than a simple re-positioning of the existing balloon space in a non-changing spatial environment. Molecules are not simply moving farther and farther away from each other unless new space has been created. And, new Space has been created. It happens everywhere all the time. The balloon. The Universe.

Cosmological Interpretation #12: Each Star Its Own Big Bang

The concept of the Big Bang, as CIG employs and extends it, is applicable to each stellar mass (star and/or galaxy cluster, etc.) as well as each atom.

One can easily recognize that there is little to distinguish the three (Big Bang, our own sun, any given atom). All three offer up their own space and time that prior to CIG, has only been applied to the Big Bang itself. The parameter property that offers up the space and time (space-time) is motion. The interplay between mass (matter) and space-time is the percentage of "c" travel. MTS

If it looks like a Big Bang, and sounds and smells like a Big Bang it is a Big Bang, no matter how small it is.

So, as you close this book, remember to go outside on a clear night and look up into the heavens. There then should be a newfound appreciation that the little specs of stars glittering in the darkness are not complacent non-entities, simply sitting there twinkle, twinkle, little star, complacent dots on an ever expanding balloon called the Universe. No, no more. The dots, the stars themselves create the expansion.

Listen and you can hear each Star Bang, and you now know that the darkness of the Dark Matter halos and, father out from the stellar center, the darkness of the Dark Energy vacuum that surrounds that star comes from that star. Each star is creating its own time and space.

If one had acute hearing sufficient to hear each Bohr orbital bang, one would be hearing each atom creating its own time and space.

The microscopic world and the macroscopic entertain the same physics. The big world is built up of the small world. The telescopes and the microscopes see the same thing.

And, since we truly don't know the scale of size except from our own perspective, we truly don't know what actual difference there exists between the scales of things. In fact, since everything is relative, we truly can only view things from a relative position, and therefore we know nothing for certain.

A subscription to CIG Theory comes with a new interpretation of the night sky.

The stars are awake.

ANNEX 1

The Fourth Law of Motion (offered today in its original format)

Synthesis of Gravity into the normal realm: Fourth Law of Motion

After having spent three full years dropping various objects onto my lab bench at "The Laboratory" as well as stretching and releasing rubber bands, etc. thinking, thinking, thinking, with associated hopes to obtain some realization as to what is happening with gravity, unified field theory, etc., the following realization materialized very distinctly and as a direct result of the fruits of those long efforts:

The following is a general outline:

Pertinent Components:

I. There is but One Particle and that I term a GODROP (Thing One).

II. All else is a part thereof and this I term GODROPLETS (Things Two).

III. Knowing Sir Newton's three Laws of Motion and his theory of Gravity.

IV. To Sir Newton's three Laws is added the following Fourth Law of Spontaneous Motion (of course it's not spontaneous—it's just the name):

FOURTH LAW OF SPONTANEOUS MOTION:

ANY accumulation of matter (i.e. pencil) occupying space and without motion with respect to its directly accompanying surrounding frame of field reference, which when allowed to be free of another holding field (the hand in this case being the holding field), and which then moves (the pencil in this case) with respect to that initial surrounding frame of field reference, are **ALL THEMSELVES ABIDING BY the SAME PHYSICAL FORCE (tendency)**. That force being and the motion of the object also being according to the following:

The Spontaneous Motion will proceed from one pressure gradient {field densities with high Time "T" value variations per unit volume}to another pressure gradient {field densities with a different Time "T" value variations per unit volume}, {and/or the same "Time" density per unit volume space but with different volumes observed}, {and/or equal volumes observed (i.e. same volume but different pressures = more time per given space!)} until such time that they (the reactive parties) reach equilibrium with one another (in this case, the pencil reaching equilibrium with the desk and now resting on the desk). The pressure difference being the relative field pressure between the two fields**, AS PERCEIVED ONE FIELD TO ANOTHER**. The field itself being the amount of matter per given volume of space (or, per MT=S, this represents the amount of Time per given volume of space). Here, the concept of pressure is being applied not only to gaseous elements but to all of matter, and in a field-like manner.

This Spontaneous Motion holds true for: objects released in air (at my desk) (Gravity), objects released in space (no movement is equal to equilibrium between perceived fields), electricity through copper wire, light through fiber optics, a rubber band stretched and released, Big Bang, metal fillings allowed to be free within a magnetic field until their movement is so stabilized, beta radiation, a burning match, a bubble of air under the water and released to allow it to "float" to the surface, the combining of elements and all of chemistry, *and*

ALL ELSE, WHEREIN PARTICLES/FIELDS/OBJECTS WITHOUT MOTION RELATIVE TO THEIR INITIAL SURROUNDING ENVIRONMENT THEN FIND MOTION, UNTIL SUCH RELATIVISTIC FIELDS OF THE ENVIRONMENT HAVE BEEN STABILIZED TO REACH EQUILIBRIUM (TIME!) AS PERCIEVED ONE FIELD TO ANOTHER. This equilibrium appears consistent with entropy, however, "*The Theory*" views entropy as a sate whereby things are moving from a state of disorder to a state of order, or equilibrium, a very ordered state, a very very ordered state. Entropy—disorder to order.

Using I and II above, along with Sir Newton's three Laws, and substituting the Law of Spontaneous Motion for Sir Newton's Law of Gravity, all things may be better understood. The four laws of motion now include gravity, radiation, etc. etc., and can be applied to everything.

Part of "FLEA" in four simple laws of motion. The concepts of gluons, electrons, protons, leptons, quarks, elements, etc. as these particles have been described as such to "explain" our world, are still simply variations on the underlying MT=S theme. In the current Standard Model, there is no end to the number of particles, simply keep simply colliding, and for that subject, why not simply extend the standard model upward and outward to the cosmic scale, where the planets themselves can be considered particles to the solar system and solar systems particles to the galaxies, for this is what the Standard Model offers, but on the micro way down, not the macro way up.

"*The Theory*" rather states that there are only GODROPLETS in a GODROP [Things 2 in THE Thing 1], following Newton's Laws 1, 2, and 3, and the above Law of Spontaneous Motion, now Law Number 4. The pressure differences explain why one particle needs an anti-particle, why a south pole needs a north pole, etc., since things are being viewed with respect one pressure gradient to another.

From the above, one can see that there are no gravitons; rather only Godroplets (Things Two) trying to reach equilibrium with each other and within the Godrop (Thing One). The Standard Model is Not Wrong, but the One Particle Aspect of *The Theory* Consolidates!

The law of gravity & the law of charges are not arbitrarily so very identical, rather they are explaining the same thing, though on different scales. *The Theory* looks at "m" and "q" not as "m" and "q" but as time-field densities. And the further these fields are apart (r^2), the less direct they "see" each other.

b) ["a" appeared elsewhere in this book, which has essentially been a cut & paste process of previous writings long ago]

On Motion Alone

All motion is simply the rate at which equilibrium is attempting to be reached. The greater the mis-equilibrium, the greater the rate of motion (the reason for catalyst in chemistry) to have the constituent particles/fields reach equilibrium. The ball dropped on the moon (as perceived by the ball to the moon) has less of a mis-equilibrium than the same ball looking at the earth, consequently, the resultant motion is also less. The greatest mis-equilibrium possible is that of the electromagnetic spectrum, wherein the rate at which equilibrium is met is the speed of light (c). My offer, though rather obscured, is to the effect, can someone create a mathematical model/formula which determines the rate at which motion will arise, when one knows the equilibrium difference between fields (mis-equilibrium). *All motion is a direct result of equilibrium being attempted to be reached (i.e. gravity, radiation, catalyst reactions, and speed of light).*

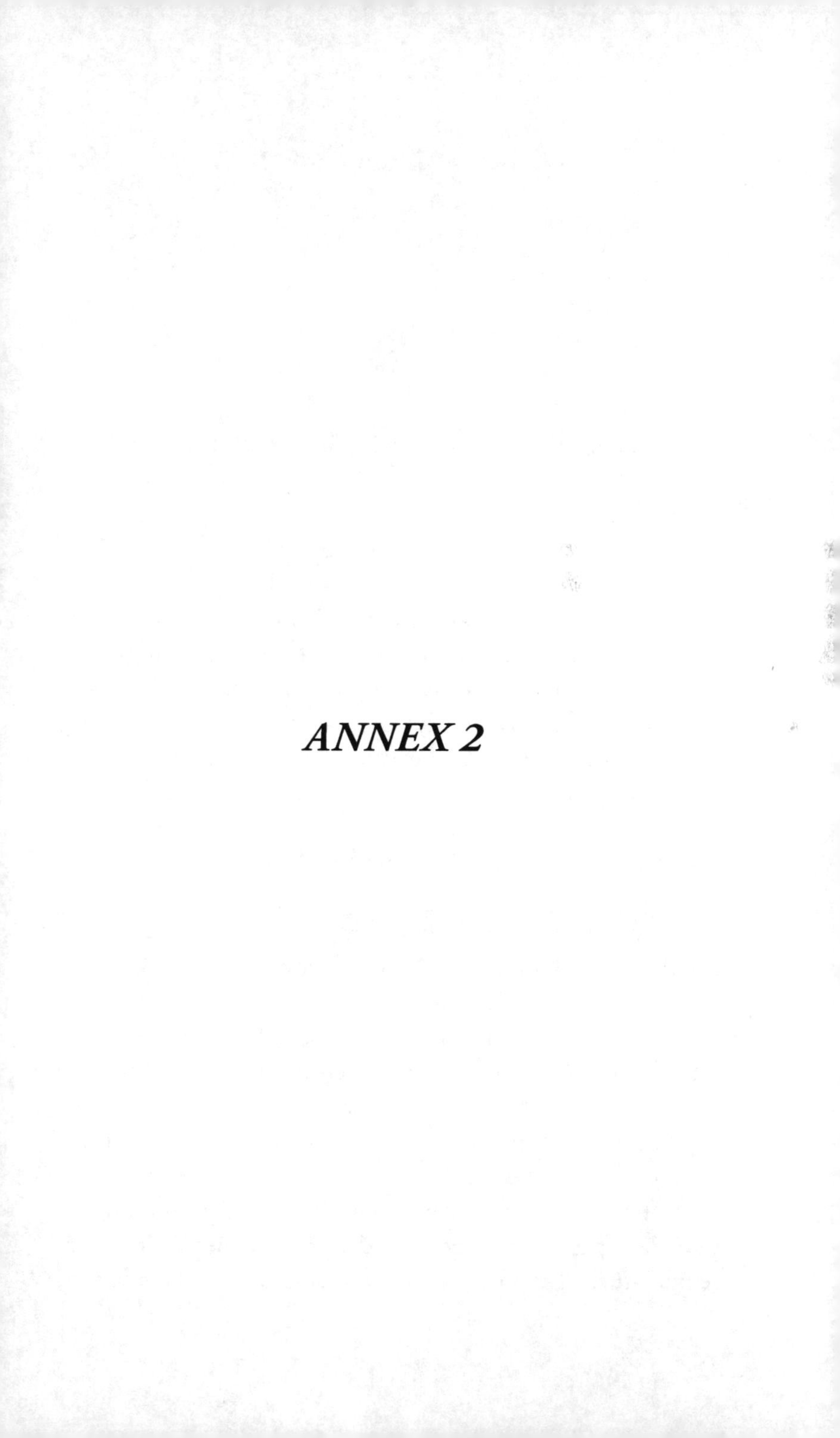

ANNEX 2

Quantum Discussion 1—

The Size

Why do large things move slower than small things?

Answer:

Larger things (i.e. planets) move slower than smaller things because unlike small faster things (i.e. electrons that turn into a more spatially manifested entity), larger things can accumulate in a more massive matter form & remain large, since they, as a slow entity, do not manifest their spatial characteristic as much as faster particles do. Slower things can be large. Faster things can't, because they remain more spatial, and hence smaller as entities unto themselves. This is why large things are slow, and small things are fast.

But of course even the largest thing is comprised of fast & small things. But unto themselves, they are slow.

So in CIG, the entire Standard Model represents various manifestations of the actual permutations of possible space-time opportunities. The space-time(s) itself becomes the particle(s) dependent upon the space-time's degree of curvature. And the rate of travel determines the degree of curvature (field density) which ultimately determines the particle's spatial size. The field density also determines the temperature of the field (recall our solution to the horizon problem). Are all identical (more or less as nothing is identical) particles (i.e. muons, photons, electrons, omega, k, sigma, pi, neutron) the same size and travel the same rate? Essentially YES—and that's what makes them the same particles . . . . Exactly, because they travel the same rate, and therefore exhibit themselves as the same size! CIG attempts to explain why the particles are the particles they are. It explains why big things are big (and slow) and little things are little (and fast).

And by now, you, the reader, should have a clear understanding of the mechanisms by which traveling mass turns to Dark Matter, Dark Energy, and of course Black Holes as you start to understand CIG Theory.

CIG also offers a reason why the particles are the particles they are.

There exists correlations with regard to particle speed versus size. If there is a Table of the Standard Model normalized to particle speed, I should like to see it, if none, then the task is on the table for the taking.

Big things are big (and slow) and little things are little (and fast), because slow things retain mass, while fast things loose mass and create space. Fast things get small as they lose mass.

Particles tend to congeal (create massively bigger entities) as things attempt to reach time equilibrium with their surrounding environment. Here, see the Fourth Law of Motion.

All massive bodies large (galaxies) and small (atoms) follow the same MTS foundational reality [dense core (tight curvature), proceeding to, respectively: heavy Dark Matter, light Dark Matter, heavy Dark Energy, light Dark Energy, pure vacuum space)].

This then because the tendency is to have each onion layering (reason for spectral lines) attempting to reach a time (movement / % "c") equilibrium with its neighboring layer.

The particles are the space-time and CIG offers that matter and space-time are one and the same. The concept that matter curves space-time and space-time tells matter how to move (as seen on TV), is understood in the theory as follows:

The space-time fabric is simply another form of the particle. MTS. And, the creation of Space is offered in the same theory.

As is the equivalency of an atomic mass unit to a spatial quantity (the CUPI quantification) at whatever rate (speed) the process took place.

The proton field density of any given periodic tabled element determines the curvature of the "electron" field surrounding it. The bigger the "proton field", the greater the electron field (number of electrons).

$E = mc^2$ is offered in CIG Theory as an actual physical concept. That being: matter to space and back to matter again [forward/reverse time vector "T" (% "c")]. All the energy is represented by matter converting to space (the blast volume in nuclear explosions) and the space back to matter.

We have digressed slightly here but it was a good digress.

Quantum Discussion 2—

The Metamorphosis

I. The Ideal Gas Law

The Metamorphosis of the Ideal Gas Law into CIG Theory's MTS Equation = a New Interpretation (adulteration!) of the Ideal Gas Law

From Wikipedia:

The ideal gas law is the equation of state of a hypothetical ideal gas. It is a good approximation to the behavior of many gases under many conditions, although it has several limitations. It was first stated by Émile Clapeyron in 1834 as a combination of Boyle's law and Charles' law. The ideal gas law is often introduced in its common form:

$PV=nRT$,

where P is the absolute pressure of the gas, V is the volume of the gas, n is the amount of substance of gas (measured in moles), T is the absolute temperature of the gas and R is the ideal, or universal, gas constant.

END Wikipedia

ENTER CIG Theory:

MTS where: M = matter and its presence resulting from the curvature of space-time (the S/T portion of the equation using vector Time), T = forward reverse vector Time and based on % "c", and S = Space itself though at various field densities/curvatures.

METAMORPHISIS

Let T (from Ideal Gas Law) = temperature = movement of the particles = % "c" = Time in the MTS equation [recall that T in CIG Theory = % "c"]

Let V = Volume (from Ideal Gas Law) = S (in CIG Theory) = Space = the new and various spatial volumes (in their respective field) created as a result of mass traveling at various rates (temperatures)

Let P = Pressure (from the ideal Gas Law) = it disappears in the MTS Cig Theory = because there is no constraint within CIG Theory (no outer boundary that restricts the vessel walls).

Let n = the amount of substance of gas (measured in moles) (from ideal Gas Law) = M = Matter in CIG Theory = the actual mass/matter/substance that has weight (is actually there)

Enter Wikipedia Again:

The gas constant (also known as the molar, universal, or ideal gas constant, denoted by the symbol R) is a physical constant which is featured in many fundamental equations in the physical sciences, such as the ideal gas law and the Nernst equation.

It is equivalent to the Boltzmann constant, but expressed in units of energy (i.e. the pressure-volume product) per temperature increment per mole (rather than energy per temperature increment per particle). The constant is also a combination of the constants from Boyle's law, Charles's law, Avogadro's law, and Gay-Lussac's law.

Physically, the gas constant is the constant of proportionality that happens to relate the energy scale in physics to the temperature scale, when a mole of particles at the stated temperature is being considered. Thus, the value of the gas constant ultimately derives from historical decisions and accidents in the setting of the energy and temperature scales, plus similar historical setting of the value of the molar scale used for the counting of particles. The last factor is not a consideration in the value of the Boltzmann constant, which does a similar job of equating linear energy and temperature scales.

END Wikipedia: Truly then I don't really understand what is happening in this regards:(i.e. The gas constant ultimately derives from historical decisions and accidents in the setting of the energy and temperature scales, plus similar historical setting of the value of the molar scale used for the counting of particles.)

REPEAT of the some of the Above:

Let T (from Ideal Gas Law) = temperature = movement of the particles = % "c" = Time in the MTS equation (recall that T in CIG Theory = % "c"

Let V = Volume (from Ideal Gas Law) = S (in CIG Theory) = Space = the new and various spatial volumes (in their respective field) created as a result of mass traveling at various rates (temperatures)

Let P = Pressure (from the ideal Gas Law) = it disappears in the MTS Cig Theory = because the is no constraint within CIG Theory (no outer boundary that restricts the vessel walls)

Let n = the amount of substance of gas (measured in moles) (from ideal Gas Law) = M = Matter in CIG Theory = the actual mass/matter/substance that has weight (is actually there)

Let R = the ideal, or universal, gas constant (from ideal Gas Law) = some type of historical correction factor = no known counterpart in CIG Theory = IGNORE "R"

SUBSTITUTION:

V = S

n = M

T = T

V = nT (new ideal gas law as morphed by CIG and as explained above)

S = MT (CIG Theory)

or, MT = S in the reverse (arbitrary direction of time = recall S/T = M was chosen to represent forward vector time = from the void in Biblical times) [in one post, using Biblical from darkness, the world was created stuff, I believe I selected 100 % "c" to zero % "c" as forward vector time]

[Recall that MTS uses both forward and reverse vector time]

cigtheory.com however still uses MT=S

so, nT = V has morphed into MT = S [please extend the imagination]

In both equations, when rate goes up (temperature), volume (space) has been created.

In MTS, since it involves rates of travel from zero to "c", space-time curvatures become apparent.

And . . . . enter CIG's explanation of Dark Matter and Dark Energy, Double Slit, etc. and the theory is vastly extended over many ranges of topics.

FUN Part:

Let Ideal Gas Law = Caterpillar

Let CIG Theory = Butterfly

II. The Einstein Field Equation

The cosmological constant Λ appears in Einstein's modified field equation. We recognize through CIG theory that it is actually a "cosmological non-constant" which varies with rate of travel (alternately temperature, space-time curvature, energy density, as these parameters are merely forms of one another).

From Einstein:

$$R_{\mu\nu} - \frac{1}{2}g_{\mu\nu}\,R + g_{\mu\nu}\Lambda = \frac{8\pi G}{c^4}T_{\mu\nu}$$

From Wikipedia:

In one field where R and g pertain to the structure of space-time, T pertains to matter and energy (thought of as affecting that structure), and G and c are conversion factors that arise from using traditional units of measurement. When Λ is zero, this reduces to the original field equation of general relativity. When T is zero, the field equation describes empty space (the vacuum).

The cosmological constant has the same effect as an intrinsic energy density of the vacuum, ρvac (and an associated pressure). In this context it is commonly defined with a proportionality factor of 8π: Λ = 8πρvac, where unit conventions of general relativity are used (otherwise factors of G and c would also appear). It is common to quote values of energy density directly, though still using the name "cosmological constant".

A positive vacuum energy density resulting from a cosmological constant implies a negative pressure, and vice versa. If the energy density is positive, the associated negative pressure will drive an accelerated expansion of empty space. (See dark energy and cosmic inflation for details.)

End Wikipedia information

<u>Enter CIG!</u>

OK—Specifically, where it is stated above, "When T is zero, the field equation describes empty space (the vacuum). & remember, "T pertains to matter and energy (thought of as affecting that structure)"[in the field equation above].

This correlates with CIG as follows:

In CIG, when matter [their T (not ours which states T=Time or % "c")] is no longer there, it has transformed into space [MT=S]. It is an interpretation that within the Einstein equation, and where T pertains to matter and energy (thought of as affecting that structure, and where it is also stated that "when T is zero, the field equation describes empty space (the vacuum)", this all correlates to CIG, whereby it is an active process through which matter manifests itself into the vacuum. It is stated many times within CIG that this is due to varying rates (% of "c"). Space-time must be broken.

The Einstein field equation already contains the CIG interpretation of nature, as explained above. But, it was never interpreted in the manner in which CIG interprets nature.

It may take two reads to figure out what I am saying, especially since the same symbol "T" represents two distinct variables [T = % "c" in my theory] and [T= matter and energy in the Great Professors equation] Get over this stumbling block and try to understand what CIG says.

III. The Combo

The below is to supplement the above posts I and II regarding the metamorphosis of the Ideal Gas Law and Einstein's Field equation into CIG Theory's MTS equation. It is to be taken in context with the Metamorphosis, and, based on substitution, the Ideal Gas law can now be substituted into Einstein's field equation as follows:

Let A = Einstein's field equation

Let B = Ideal Gas Law

Let C = CIG Theory

Posts I & II correlated A = C and B = C

By substitution A = B

Therefore Conceptually:

A = Einstein's field equation = B = Ideal Gas Law = C = CIG Theory

[Again, please extend the imagination]

ANNEX 3

The Ramblings

We have looked into the relationship between Time, Space and Matter. We have looked at the root of Matter, and found its tangible components to consist of Space and Time. The title "*I Have Become Space*" is now ours to consider.

And the next time we look up at the heaven's stars, we see that the immediate space surrounding each star is still being created from the stars' mass. The matter unfolds into it becoming the space itself, this as the speed of light emanating from the star reaches "c" value.

As the first percentages of the speed of light is reached, the dark matter (more massive) is neatly unfolded, with the dark energy (pure space) all unfolding at the greater speeds approaching "c". Through this reasoning we know why Dark Matter Halos appear nearer to massive bodies and not in pure open space. That reason being is that the massive photons shed less mass at the beginning of their journey (i.e. before the mass reaches full "c" value"), and retains some value of matter, as "dark matter".

During light's journey, the light fights the gravitational pull of the star it emanates from. In cruder terms, it "sticks around for a while" in its slow form, thus the Dark Matter is created. *The Theory* dictates that these unfolding's follow the percentages of Lorentz transformation equations, of time dilations. As length contracts, space unfolds. Most of the spatial volume of the current "Our Universe" [as opposed to the philosophized "Many Universes" or "Parallel

Universes"] was created during a Big Bang. But space still casually unfolds today in stars and atoms and anything that moves, through the de-materialization process described as "MT=S". And through the materialization of our "space-time-matter-light-microphotonic-structural-macrogrowth-wave-field-theory", through the process "S/T=M", space also again becomes matter, and it is this most common occurrence at the micro-scale, upon which the macro is built.

It also appears that the Space /Time/ Matter permutations/ combinations are sufficiently large enough so as to allow for the multitude of realities that we consciously see around us.

This is your paradigm shift.

The probability that separate and distinct particles (electrons, protons, neutrons, quarks, leptons, neutrinos, anti-neutrinos, photons, etc.) exist solely and exclusively by themselves, retain their own distinct properties, and have, on their own accord, "found" their way into each and every atom—electrons around the outer orbit, protons / neutrons in the nucleus core is simply "Ridiculous". The probabilities are just too astronomical.

Absolutely "Ridiculous"!

Great for explaining things, mathematically & fundamentally correct when determining outcomes (i.e. chemistry, fusion, etc.), but the questions remains: Is there a simpler way to examine and explain the universe, both on a macro and micro level? A one-particle view of nature using a view of entropy in which "*time equilibrium*" is the motivating factor is inherent in *The Theory.*

CIG rather states that "everything is the same" (Defined in *The Fourth Law* as Godrop or Thing One). One particle only! As close to God as physics, with all its metaphysical origins, and in the context of this paper, may reach.

The next step may be God, fusing theology, philosophy, religion, and physics, as one, as God, and for the atheists, as nothing, for nothing and pure being are one and the same, as a past philosopher notes.

That the constituent subunits of the Godrop are "godroplets" {if you remain averse to risk, and do not wish to view *The Theory* in terms of the Godrop and godroplets, at your option simply feel free to substitute Thing One and Things Two} defined for the purpose of introducing the necessary means to explain the interaction of these sub-units, as one unit reaches Time Equilibrium with another. That based on the relative nature of these "same any particles" {(godroplets) (Theory defines particles of the Godrop as godroplets, since we are to break down the single one and only Godrop into its fundamental and many sub-component godroplets for the purpose of taking apart/understanding our world)} as they are relative to one another "same any particles" (godroplets = things two) actually determines what the overall property outcome will be. That outcome is determined by the Fourth Law of Motion. This new law allows us to consolidate all of the known particles into these subunits of the single Godrop; these subunits are the godroplets.

Consider Sir Newton's first three laws and then his Law of Gravity—appearing to hang away by itself. Then consider the lack of a law that explains "spontaneous motion" (as defined in theory). Then recognize that the fundamental nature of movement (from no movement) is so much so at the very root of nature, and therefore the science of Physics overall, that it must be the result of a single phenomenon. That to find motion from no motion is at the root of nature, and can be explained from a single law. This philosophy parallels that philosophy wherein "an object at rest shall remain at rest and an object in uniform motion in a straight line shall maintain that motion unless acted upon by an outside force". This new law equates the motion of metal filings near a magnet, to gravity, to the reason chemistry takes place, to fusion, radiation, in fact, to motion itself.

To help understand these concepts this technical paper attempted to approach the subject in an easy step-by-step manner.

Particles tend to congeal (create massively bigger entities) as things attempt to reach time equilibrium with their surrounding environment. This is gravity. When there exists a disequilibrium, motion follows the Fourth Law. The motion inherent to radiation and a falling object due to gravity are equated in the Fourth Law.

The measurement problem is not a measurement problem in CIG. To measure, one has to stop the event, and once stopped, there is a collapse of the space. This then is what occurs during the measurement. You would need to video the event at the same rate it occurs to obtain the event without collapsing the event. Only in this manner would each frame realize the spatial existence of the particle. [a regular video doesn't work].

All massive bodies large (galaxies) and small (atoms) follow the same MTS reality [dense core (tight curvature), proceeding to, respectively: heavy Dark Matter, light Dark Matter, heavy Dark Energy, light Dark Energy, pure vacuum space)].

This then because the tendency is to have each onion layering (reason for spectral lines) attempting to reach a time (movement/% "c"") equilibrium with its neighboring layer.

The particles are the space-time and CIG offers that matter and space-time are one and the same. And time is movement and this obviously correlates to temperature. So temperature and time and curvature and space and matter all become simply a form of one another. The concept that matter curves space-time and space-time tells matter how to move is understood as follows:

This is Quantum Gravity and CIG combines the fabric and the mass.

For equivalencies of mass units to spatial quantities use the CUPI quantification and whatever rate (speed) the process took place.

The proton field density of any given periodic tabled element determines the curvature of the "electron" field surrounding it. The bigger the "proton field", the greater the electron field (number of electrons). The desire to reach time equilibrium is the driving means.

$E = mc^2$ is offered in CIG Theory as an actual physical concept. That being: matter to space and back to matter again (forward/reverse time vector "T" [c]). All the energy is represented by matter converted to space (the blast volume in nuclear explosions) and the space back to matter.

Please understand CIG Theory.

The Venturi Rambling:

The Venturi principle comes into play in CIG—the rate at which something flows creates different pressure gradients—the higher the rate, the lower the pressure:

Wikipedia

The Venturi effect is the reduction in fluid pressure that results when a fluid flows through a constricted section of pipe. The fluid velocity must increase through the constriction to satisfy the equation of continuity, while its pressure must decrease due to conservation of energy: the gain in kinetic energy is balanced by a drop in pressure or a pressure gradient force. An equation for the drop in pressure due to venturi effect may be derived from a combination of Bernoulli's principle and the equation of continuity.

End Wikipedia

Enter CIG

It appears too coincidental that the high rate at which light proceeds, and the resultant space that manifests from that (MT=S), creates the extremely low pressure of Space (i.e.

pure vacuum), from the highest rate possible, that of "c". Highest rate transforms to lowest pressure. A corresponding attempt at some correlation: Venturi = CIG

A light rambling on the correspondence principle:

The great Neil Bohr's correspondence principle demands that classical physics and quantum physics give the same answer when the systems become *large*.

CIG offers the following: Bohr's correspondence principle demands that classical physics and quantum physics give the same answer when the systems become *slow*.

More precisely:

The purely classical, or that matter which does not follow De Broglie's wave-particle duality is found only at zero rate (zero "c"). Anything (matter) traveling at any speed will experience quantum (diffraction).

The conditions under which quantum and classical physics agree are referred to as the correspondence limit, or the classical limit. Therefore, technically, quantum and classical agree at the junction of zero mph and that smallest rate of travel (unit of temperature) possible above zero mph. I do not know the smallest increment of rate travel, but let's designate it a *sloth* if one does not yet exist in the books. One sloth over zero = quantum.

Since we are not used to this (i.e. thinking of ourselves as quantum), we must then arbitrarily define a point where classical and quantum give the same answer in terms not of large, but of speed. Use the Lorentz curve for time dilation or length contraction, and pick a point on the curve. We then arbitrarily define the classical limit, as that point beyond which if the matter travels faster it is no longer classical but now resides in the quantum realm. It is herein offered that point be designated the "Arikian" and that it represents the agreed upon point that classical physics and quantum physics give the same answer. It is a point of convenience, otherwise we are stuck with a black hole being the only classical thing in town. Who will agree on that point?

How do we experimentally verify CIG Theory?

Perhaps through fair analysis of Cosmological data of Red Shift Anomalies, the balloon in the refrigerator experiment* followed by deep philosophical discussion, the blast energy of a nuclear bomb, as calculated using the CUPI quantification. Just suggestions.

* A small request to those inclined:

CIG theory introduces a new science of pressure (newly created space based on traveling massive particles).

Could the mathematicians calculate that it [ ] is; [x] is not possible for, in our refrigerator experiment, a deflated balloon to enlarge when heated, based on the current view of particle bombardment against the balloon walls as the sole cause for expansion. Does the particle mass and acceleration (F=ma), when taken in its accumulated form, create enough force to actually press outward on the balloon's wall, to inflate it? Please take a worst case scenario (e.g. a heavy walled balloon). My guess is it doesn't, and therefore another explanation would be necessary to explain the phenomenon (Enter CIG theory).

I know that there is a lot of math involved, and, for professors, maybe this could be for your students, as a project. Perhaps give it to three separate groups of students, and compare their independent results.

Please crunch the numbers.

A point confirmation of CIG Theory is at stake.

Annalen der Physik: A noble attempt to publish my theory in a most respected Journal.

Subject: AdP—Decision on Manuscript ID andp.201200046

07-Mar-2012

Dear Mr. Lipp,

I write you in regards to Manuscript ID andp.201200046 entitled "The Coney Island Green Theory" which you submitted to Annalen der Physik.

I am sorry to tell you that your manuscript will not be further considered for publication in Annalen der Physik. Since we receive many more manuscripts than we can possibly publish we have to focus on those with the most overlap to our scope. Unfortunately your manuscript does not qualify in a formal, nor in a technical sense for our journal.

Thank you for considering Annalen der Physik for the publication of your research.

Sincerely,

Editor, Annalen der Physik

CIG response:

Subject: Re: AdP—Decision on Manuscript ID andp.201200046

Dear (Annalen der Physik)

Thank you kindly for the most prompt of replies. I do hope that you have taken enough time to read and actually understand the theory and its offerings. If so, then it should be recognized that your notation to "most overlap to our scope" does not quit fit, since CIG literally offers a new physics.

However, I am most agreeable to your comment as regards CIG qualifying "in a formal sense", because it obviously does not. There is full concurrence here. But, what CIG Theory lacks in a formal sense is however fully made up for in a technical sense. Naturally, some technical aspects are of course lacking (aren't we all), but what is not lacking is the overall true description of nature, that is, Matter actually turning into newly created Space.

I agree to respectfully disagree on matters of a technical sense. I do hope you will make another attempt at understanding the theory and will reconsider your current position.

Thank You,

Douglas William Lipp

Now, for those of you who may not know, Annalen der Physik was the Journal that Albert Einstein published his famous papers in. It is therefore the highly esteemed journal and I thought, hey why not reach for stars.

Back to earth . . .

Any theory that displaces the current norms of Physics and Cosmology to the degree that CIG does is facing an arduous road of acceptance:

"All truth passes through three stages. First, it is ridiculed. Second, it is violently opposed. Third, it is accepted as being self-evident."

—German philosopher Arthur Schopenhauer (1788-1860)

Now, perhaps the theory is correct, near correct, or by all appearances, completely flawed from every angle.

Better the theory be wrong and exist for contemplation, than not exist at all, never to be contemplated.

However it should also be noted that the more things that can be explained by a theory, the more apt that theory to be correct. One day perhaps CIG will be self-evident.

"All truth passes through Three Stooges. First, it is ridiculed by Moe. Second, it is violently opposed by Larry. Third, it is accepted as being self-evident by Curly."

—Unknown American philosopher (2014)

A preempt of the anticipated attack on units:

Dear ______,

Thank you kindly for your reply.

The matter of units is a topic I worked on briefly, as this is the second time it has been offered. It appears that the application of units is one of the first things that physicists use to assess whether an equation has flaws. I never quite resolved the units issue as it was easier to dismiss. I dismissed it, as follows: As an analogy, if I am thinking correctly (remember math is not my expertise, even simple as this is) prior to $E=mc^2$ (J=g m/s m/s), I don't believe that grams (mass) could have been taken into unit agreement with Joules (energy). Is there unit agreement in $E=mc^2$? Was there before the equation?

Also, as for example if apples were always known as apples and pears as pears, and oranges as oranges, and were they known units, and then some theory comes along and shows that there is an equivalency between them such that, in terms of units apples could be understood as oranges divided by pears, then so the new thinking (with much resistance of course) would be that apples is oranges divided by pears. I offer that there is a spatial equivalency (cubic meters) to

Mass (grams), and so, prior unit agreement or not, we must now accept the new conversion.

I thank you for bringing these comments forward. It allows me to focus on my rationale for my theory, and it is exactly what I need. Now I know that there are so many questions I will not be able to answer because the physics is beyond me, and my math is pathetic. Conceptually though, and rationally, I believe I can fully defend my theory and that it will hold up to the rigor of experimentation as well.

I was very tempted to simply say "no caramel apples for you"!

End—Unit Agreement

Start: The vacuum catastrophe

(Wiki)—In cosmology the vacuum catastrophe refers to a disagreement of 107 orders of magnitude between the upper bound upon the vacuum energy density as inferred from data obtained from the *Voyager* spacecraft of less than 10^{14} GeV/m^3 and the zero-point energy of 10^{121} GeV/m^3 suggested by an application of quantum field theory.

(CIG) It may just be possible, though perhaps very unlikely, that CIG can offer some insight into this "catastrophe". For instance, might quantum field theory be missing the fact that matter is becoming space; that the actual recorded value of the energy density is being diminished by the new MTS volumetric increase, thereby making the inferred data from *Voyager* much less than calculated quantum field theory data? Who knows? Perhaps quantum field theory renormalized to include new CIG volumes, will result in something closer to 10^{14} GeV/m^3 for the vacuum energy density.

END THE RAMBLINGS

Annex 4: The Distance Equation

As we all knew before reading *"I Have Become Space"*, Rate multiplied by Time = Distance

But now that you know that there is another component to that equation.

That RT = D where: R = rate, T = time, and D = distance is well known.

However, in the equation that we are looking for, RT is only one component. To contemplate the "new space" (distance) resulting from traveling mass, another component that will be some function of mass and the rate at which it travels (follow Lorentz transformation for time dilation) will be needed. This second component of the equation will represent the new distance supplied by the moving particle. The CUPI quantification is the equivalency of one atomic mass unit as it travels at rate "c".

0.02762u = 25.7MeV= 14,952,942.08 pm cubed of space
(Mass) (Energy) (Space)

1u = 541,380,958.7 pico meters cubed of space (at rate "c")

Rate x Time = Distance (near true, classically for slow moving particles, as per CIG Theory)

That gets us from starting point A to a Point B.

But, in an expanding universe, Point B is now a farther away Point C. In other words, the current distance equation of RT must be modified to account for the new distance. CIG theory offers that the new distance is the result of the moving particle itself. And, the final accelerating Universe is the result of the accumulation of these individual expansions.

We recognize that rate x time can't work.

Further, CIG theory offers that each moving particle (in this case let's limit the concept to stellar photons that start with mass, and offer up that mass by manifesting into their spatial volumetric equivalencies) (this solves Red Shift anomalies).

So, CIG offers that the overall Universal expansion (think Red Shift & Type 1A supernovas) is the result of each stellar entity adding their own bit of volumetric space (also offered for new Bohr orbitals, and anytime a massive particle moves).

And, we are looking for some sort of mathematical component to account for what may be a compounding effect of these "moving particles" adding up one on top of the other.

I am not a mathematician and so can only offer thoughts, not the equation itself.

But, to help find it, here would be my quick answer:

The equation should be interpreted as:

The variables of the current distance equation R x T will therefore be supplemented by those variables that will compensate for this new expansion of space. As you may be aware, CIG Theory offers that as a piece of matter moves, its offers up a certain new spatial volume, at the expense of mass. (therefore, CIG Theory is at odds with Einstein's view that mass increases as it moves faster and faster, and increases to infinity at the speed of light; this is not true, mass actually decreases as it moves faster and faster).

OK so how shall we begin?

First, the R x T is the first component of our equation, since this still holds true. We are adding to that: the new CIG component.

R x T + "???"

"???" = CIG component necessary to define true distance equation

I am therefore going to explain the "???"

GIVEN:

1. given

"???" will contemplate the transformation of mass to space at the following equivalency and at the speed of light (internal note: assumed that the fusion process takes place at this rate):

lu = 541,380,958.7 pico meters cubed new space [this new distance (in unidirectional meters) is the new distance that we are adding to correct the current RT=D equation].

u = one atomic mass unit (1/12 of the mass of the carbon atom)

We then re-normalize this "one atomic mass unit" to the gram.

I'm not sure what unit will work best (mg, g, Mg, etc.) as the new offered space works on the stellar scale (the space between stars) as well as the micro (new space of new Bohr orbitals).

2. given

The transformation of the gram unit (whatever is decided) to its equivalent spatial volume, follows Lorentz Transformation as that for Time Dilation. Further reading see:

http://en.Wikipediapedia.org/Wikipedia/Lorentz_transformations

The equation that we will be incorporating into ours can be found somewhere at:

http://en.Wikipediapedia.org/Wikipedia/Time_dilation

I believe it is:

1 / the square root of 1 minus "v" squared/ "c" squared

Where, v = velocity and c = the speed of light

And, this is for a constant acceleration

If you want to try the equation to represent a non-constant acceleration, this gets rather involved (let's start with constant acceleration first)

3. given

So, mass will transform (offer up its equivalent spatial quantity) at the same percentages as time dilation correlates to % speed of light. And, in volumes that correspond to the equivalency above (i.e. 1u = 541,380,958.7 pico meters cubed)

4. given

Since each stellar mass turns to space (the equivalency), we have to sum the distances. Therefore, our equation has to have some sort of summation process to it. (This plus this plus that, etc.) This will account for each mass entity creating its own space (metered distance for our purposes in our equation). Combined, this represents the accelerating Universe.

[Internal note: Charles law and Boyles law represent new space and not the old understanding of particles moving faster and faster and simply away from each other without the creation of new space] [In this regard CIG re-interprets Charles and Boyles Law which now foundationally support CIG Theory]

So, there will need to be a summation component as well.

5. given

R x T (current component) + mass [turning into metered unidirectional spatial measurement using % time dilation & cig transformations] + a math component representing the accumulative summation of more than one moving mass = our new and True Distance equation.

We were looking for an equation that adds that little extra distance the moving particle itself creates, and to the nth degree when many particles are involved as in the accelerating Universe. All this since the particle finds itself a little farther away than it classically would under RT.

We have not found our equation. We want our equation!

The Vatican

Link to Biblical Years

No theory of this likes, void of heresy, is complete without some Biblical synergy to consider, perhaps by the Vatican itself, since that entity has been so historically involved in furthering inquisitive knowledge.

For consideration among other things so desired:

Enter the Biblical time bubble:

If the spatial bubbles (each star its own Big Bang) considered herein are cumulative, and do not represent a straight line single One Big Bang process to the end of the Universe, rather, use the many "bubbles upon bubbles"— "additive in nature"- "Coney Island Green Theory" interpretation to obtain the end resultant "One Giant Universal Bubble", and if we "abrogate" for a lack of better word, these little stellar bubbles of Dark Matter and Dark Energy, reeling them in like many fishermen reeling in many fishing lines (as opposed to the current "one fishing one" theory), do we not end up with a time nearer to the Biblical years as opposed to billions of years?

Perhaps using the theory's "light based matter" turned into "many individual bubbles of space" view.., if we pull back all bubbles, do we end up with biblical years? I.e. 6000 years? If each bubble were additive how would this new time dimension be expressed; how many years would we have?

To quote Brian J. Hatton, "Me and you are the same as the Rock, or the Seagull, or the Dolphin. It's all a matter of the time & space that we are within".

I believe Peter was the Rock ["And I also say to you that you are Peter, and upon this rock I will build My church; and the gates of Hades shall not overpower it," (Matt. 16:18)].

And the matter is the time and space.

Thank you for considering CIG Theory. Keep an open mind but don't be stupid. To fire breathing dragons and conspiracy theories, to string theory, CIG Theory, golden sunken treasures gilded with guilt, and infinity and beyond, to quenching deep thirst with sips of cool water, to the sublime less is more, the very small, to the Cosmopolitan, we now must say goodbye.

Well hello there. Beautiful night sky. Have you heard of CIG Theory?

(OK, now . . . spill the drink)

THE END

DEDICATION & ACKNOWLEDEGEMENTS

Dedicated to my dearest wife of 31 years, Lian Siang O Lipp (the original CUPI), the girl who literally keeps me alive. And to Forrest and Trevor, Jacklyn and Sabrina, the next generation of thinkers.

Also dedicated to all the Physicists and Cosmologists of both worlds. Let's meet in the middle.

And to Professor Faramarz Ghassemi, who wrote a paper delving into the concept of protons turning to space. And gave me encouragement. And to Dr. C. Sütterlin of Jean de Climont Associates Ltd who allowed me to be a real dissident. Thank you.

Special thanks to Ed Arikian throughout the years. Ed knows.

And to all those who have shown interest. Giacomo, Peter, where are you? Peter—you know my theory better than I do.

To Wikipedia, TV, NOVA, the Closer to the Truth videos, and FQXi, and all the other places where I attempted to learn these topics. No homework!

And to Fred Bowditch who gave me the $2.98 Bantum book: The Universe and Dr. Einstein by Lincoln Barnett, Foreword by Albert Einstein, in Santa Barbara long ago.

And most certainly to my dad who kept the record: Edward Teller— "The Size and Nature of the Universe / The Theory of Relativity" out and about when I was just a kid. Side 2 was tough listening. He showed me where the stars were. Look up.

And more most certainly to Mom. And to my brother David Scott Lipp and his wife Terri.

The Apologies

Finally, my deepest apologies for the randomness and nuisance of my emails to the community and for the utter lack of use of the scientific method.

This book relegates itself to the time honored art of extending universal knowledge

The Coney Island Glossary

The following glossary, offered at times with compliments to Wiki, is for the non-scientist, myself included, to familiarize oneself with some conceptual ideas presented in the book.

Accelerating Expansion of the Universe—The concept that the Universe is not only expanding, but the expansion is accelerating. Faster, faster . . .

Big Bang—The cosmological model for the development of the universe and where it started in a hot and dense state, then expanded rapidly, and finally cooled into today. CIG offers that stars and atoms are their own Big Bangs.

Black Holes—A state of matter so dense and with gravitational pull so strong, that light cannot escape it. In this book, the "M" singularity side of the MTS equation.

Bohr Orbitals—An atomic or Bohr model orbital describes the wave-like behavior of electrons in atoms. The model can be used to calculate the probability of finding any electron of an atom in any specific region around the atom's nucleus. As CIG explains, the electron is a field when moving and collapses to a point particle when stopped. New space is created that wasn't there before whenever an electron jumps to a higher orbital. There is no probability as the electron is spatially everywhere CUPI units of volume allows it to be.

"c" : The speed of light in vacuum. Its value is 299,792,458 meters per second.

Conservation of Energy (Law of)—The concept that Energy can neither be created nor destroyed but only changes state. CIG offers that mass changes to its equivalent energy state in the form of space through the workings of the MTS equation. It explains how there can exist vacuum energy & an accelerating expansion of the Universe while still adhering to the law.

Cosmological Constant—That value added by Einstein in his field equation to counter the pull of gravity in order to maintain the static Universe he believed in. His correct blunder. CIG offers "rate driven Cosmological non-constants".

Double Slit—The double-slit experiment is a demonstration that light and matter can display characteristics of both classically defined waves and particles. Thomas Young (1773-1829) first demonstrated this phenomenon. As CIG explains, the particle is classically a particle only when it has no movement. Some fun videos on the topic can be found on You Tube. Hint: look for the cartoon Professor. The Double Slit is the granddaddy of all quantum experiments.

Decoherence—In quantum mechanics, quantum decoherence is the loss of coherence or ordering of the phase angles between the components of a system in a quantum superposition. Quantum decoherence gives the appearance of wave function collapse (the reduction of the physical possibilities into a single possibility as seen by an observer) and justifies the framework and intuition of classical physics as an acceptable approximation: decoherence is the mechanism by which the classical limit emerges from a quantum starting point and it determines the location of the quantum-classical boundary (The Arikian Point). In CIG Theory, it is viewed as the actual collapse into the more point particle phase when the particle slows down. In CIG Theory, the "physical possibilities" noted above are viewed as the new spatial reality per the MTS equation and the CUPI.

Dark Matter—Invisible gravitationally pulling matter.

Dark Energy—Invisible pushing or expanding matter.

Entanglement—Entangled with non-locality (see non-locality this glossary).

Horizon Problem—The horizon problem points out that different regions of the universe cannot possibly contact each other because the distances between them (horizon to horizon) exceed the "limits of time contact" and therefore must not have the same temperature. But they do, and that's the problem. CIG unfolds temperatures locally.

Mass-Energy Equation ($E=mc^2$) : Einstein's famous equation. Where: E = energy, m = mass, and c^2 = speed of light multiplied by itself

Measurement Problem—In quantum mechanics, wave function collapse is the phenomenon in which the superposition of several eigenstates appears to collapse to a single state after being measured. Hence the process of measurement connects the wave function with classical observables like position and momentum and "does something" to the environment. CIG explains rather simply the reality behind this "measurement problem". It explains the "does something".

Non-Locality—Non-locality or action at a distance is the direct interaction of two objects that are separated in space with no perceivable intermediate agency or mechanism. Quantum non-locality refers to what Einstein called the "spooky action at a distance" of quantum entanglement. CIG does not subscribe to either non-locality or entanglement.

Quantum Gravity—Quantum gravity (QG) is a field of theoretical physics that seeks to describe the force of gravity according to the principles of quantum mechanics. CIG claims to have done this by offering that gravity (space-time curvature) becomes matter (the quantum world of the Standard Model).

Quantum Tunneling—Quantum tunneling refers to the quantum mechanical phenomenon where a particle tunnels through a barrier that it classically could not surmount. It explains how the chocolate chip cookies on the table all of a sudden disappear right through the table. Or why you can walk through walls (if you travel at or near the speed of light). As CIG explains, the particle when in its spatial form, has no problem surmounting a classical barrier.

Red Shift & Red Shift Anomalies—Redshift occurs when wavelengths are shifted to the red end of the electromagnetic spectrum. These shifts are accompanied by an increase in wavelength, lower frequency, and lower energy. A redshift occurs whenever a light source moves away from an observer. Cosmologically speaking, this offers the rationale for expansion of the Universe. Using concepts of the Big Bang, this redshift should be uniform. However, observations of astronomical stellar objects reveal non-uniformities in the redshift. These are called anomalies. CIG explains away these anomalies.

Sonoluminesence—Is the emission of short bursts of light from imploding bubbles in a liquid when excited by sound. CIG explains the phenomenon as S/T = M.

Space-time: The fabric upon which matter takes to the stage. The combination of the three familiar dimensions of space with the one dimension of time. The single continuum of space and time.

Superposition—Quantum superposition is a fundamental principle of quantum mechanics that holds that a physical system—such as an electron—exists partly in all its particular theoretically possible states & configurations simultaneously. But, when measured or observed, it gives a result corresponding to only one of the possible configurations. CIG offers that the "theoretically possible states" are in reality the new spatial form of the particle.

Uncertainty Principle—In quantum mechanics, the uncertainty principle asserts a fundamental limit to the precision with which certain pairs of physical properties of a particle known as complementary variables, such as position x and momentum p, can be known simultaneously. The physicist Werner Heisenberg stated that the more precisely the position of some particle is determined, the less precisely its momentum can be known. CIG offers the Certainty Principle.

Virtual Particles—In physics, a virtual particle is a transient fluctuation that exhibits many of the characteristics of an ordinary particle, but that exists for a limited time. Virtual particles are excitations of the underlying vacuum fields and exist temporarily. CIG believes they may arise from the collapse of the vacuum field, matter from space, in accordance with the MTS equation. In the case of space to matter, this is S/T =M.

Wave Function (collapse of)—A wave function is a mathematical tool to describe the quantum state of a particle. It helps to describe how the system behaves and is a central concept in quantum mechanics. The wave function represents the probability amplitude of finding the system in that state. It gives rise to the wave-particle duality discussed in the Double Slit experiment. It appears though only to offer a "probability" of finding a particle in a given place at a given time, and this, only if the particle's position is measured. The measurement itself collapses the wave function to a particular place in time and space.

Z Theory—A sleepy theory. One which may have potential but has not awakened in the mainstream community. CIG theory is a Z theory.